Statistics for Social and Behavioral Sciences

Series editor
Stephen E. Fienberg
Carnegie Mellon University
Pittsburgh
Pennsylvania
USA

Statistics for Social and Behavioral Sciences (SSBS) includes monographs and advanced textbooks relating to education, psychology, sociology, political science, public policy, and law.

More information about this series at http://www.springer.com/series/3463

Frank B. Baker • Seock-Ho Kim

The Basics of Item Response Theory Using R

Springer

Frank B. Baker
Educational Psychology
University of Wisconsin-Madison
Madison, WI, USA

Seock-Ho Kim
Educational Psychology
University of Georgia
Athens, GA, USA

ISSN 2199-7357   ISSN 2199-7365   (electronic)
Statistics for Social and Behavioral Sciences
ISBN 978-3-319-54204-1   ISBN 978-3-319-54205-8   (eBook)
DOI 10.1007/978-3-319-54205-8

Library of Congress Control Number: 2017933685

© Springer International Publishing AG 2017
This work is subject to copyright. All rights are reserved by the Publisher, whether the whole or part of the material is concerned, specifically the rights of translation, reprinting, reuse of illustrations, recitation, broadcasting, reproduction on microfilms or in any other physical way, and transmission or information storage and retrieval, electronic adaptation, computer software, or by similar or dissimilar methodology now known or hereafter developed.
The use of general descriptive names, registered names, trademarks, service marks, etc. in this publication does not imply, even in the absence of a specific statement, that such names are exempt from the relevant protective laws and regulations and therefore free for general use.
The publisher, the authors and the editors are safe to assume that the advice and information in this book are believed to be true and accurate at the date of publication. Neither the publisher nor the authors or the editors give a warranty, express or implied, with respect to the material contained herein or for any errors or omissions that may have been made. The publisher remains neutral with regard to jurisdictional claims in published maps and institutional affiliations.

Printed on acid-free paper

This Springer imprint is published by Springer Nature
The registered company is Springer International Publishing AG
The registered company address is: Gewerbestrasse 11, 6330 Cham, Switzerland

Preface

This book is an update of the original book, *The Basics of Item Response Theory*, by the first author. The original book by Frank B. Baker was based on the course given during his tenure at the University of Wisconsin–Madison. It appeared in 1985. The second edition of the original book by Frank B. Baker appeared in 2001 with a publisher's note by Lawrence A. Rudner. About 15 years have passed since the last revision. So much has happened in the fields of educational measurement and psychometrics as well as in the statistical computing technology. In the meantime, we felt that the usefulness of the book would be increased by some further changes. The main alterations are due to the use of the computing package R for the illustration purpose and especially for the computer sessions. The treatment of the original topics over eight chapters has not been changed.

The original object of the book was to make the book to be a tutorial for item response theory suited to those who possess only a limited knowledge of educational measurement and psychometrics. We have never lost sight of such an object. The amendments in this book are not due to any alteration in the original object but they are necessitated by the development of the statistical computing technology. In particular, the book now aims at covering both the basics of item response theory and the use of R for preparing graphical presentation in the item response theory related writings.

We will be indebted to any reader who calls our attention to errors or obscurities.

Madison, WI, USA	Frank B. Baker
Athens, GA, USA	Seock-Ho Kim
January 2017	

Acknowledgments of the Original Book (1985)

Over the past century, many people have contributed to the development of item response theory. Three persons deserve special recognition. D. N. Lawley of the University of Edinburgh published a paper in 1943 showing that many of the constructs of classical test theory could be expressed in terms of parameters of the item characteristic curve. This paper marks the beginning of item response theory as a measurement theory. Dr. F. M. Lord of the Educational Testing Service has been the driving force behind both the development of theory and its application for the past 35 years. Over this period, he has systematically defined, expanded, and explored the theory as well as developed the computer programs needed to put the theory into practice. This effort culminated in his recent book on the practical applications of item response theory. In the late 1960s, Dr. B. D. Wright of the University of Chicago recognized the importance of the measurement work by the Danish mathematician Georg Rasch. Since that time, he has played a key role in bringing item response theory, the Rasch model in particular, to the attention of practitioners. Without the work of these three individuals, the level of development of item response theory would not be where it is today.

I am indebted to Mr. T. Seavey of Heinemann Educational Books for first suggesting that I do a small book on item response theory. This suggestion allowed me to fulfill a long-standing desire to develop an instructional software package dealing with item response theory for a microcomputer. I must also acknowledge the technical assistance of Mr. W. Vilberg in squeezing the maximum capability out of the Apple II computer. Without his help, the computer software would be much less sophisticated. Finally, the manuscript was prepared using the Screenwriter II word processor program written by Mr. R. Kidwell of Sierra On-Line, Inc. Without this marvelous package, the present book would never be written.

Madison, WI, USA   Frank B. Baker

Acknowledgements of the Second Edition of the Original Book (2001)

Over the past century, many people have contributed to the development of item response theory. Three persons deserve special recognition. D. N. Lawley of the University of Edinburgh published a paper in 1943 showing that many of the constructs of classical test theory could be expressed in terms of parameters of the item characteristic curve. This paper marks the beginning of item response theory as a measurement theory. The work of Dr. F. M. Lord of the Educational Testing Service has been the driving force behind both the development of theory and its application for the past 50 years. Dr. Lord systematically defined, expanded, and explored the theory as well as developed the computer programs needed to put the theory into practice. This effort culminated in his classic books (with Dr. Melvin Novick, 1968; 1980) on the practical applications of item response theory. In the late 1960s, Dr. B. D. Wright of the University of Chicago recognized the importance of the measurement work by the Danish mathematician Georg Rasch. Since that time, he has played a key role in bringing item response theory, the Rasch model in particular, to the attention of practitioners. Without the work of these three individuals, the level of development of item response theory would not be where it is today.

I am indebted to Mr. T. Seavey of Heinemann Educational Books for first suggesting that I do a small book on item response theory, which resulted in the first edition of this book in 1985. This suggestion allowed me to fulfill a long-standing desire to develop an instructional software package dealing with item response theory for the then-state-of-the-art Apple II and IBM PC computers. An upgraded version of this software has now been made available on the World Wide Web (http://wricae.net/irt).

Madison, WI, USA   Frank B. Baker

Contents

1. The Item Characteristic Curve 1
2. Item Characteristic Curve Models 17
3. Estimating Item Parameters 35
4. The Test Characteristic Curve 55
5. Estimating an Examinee's Ability 69
6. The Information Function 89
7. Test Calibration 105
8. Specifying the Characteristics of a Test 127

A. R Introduction 137

B. Estimating Item Parameters Under the Two-Parameter Model with Logistic Regression 165

C. Putting the Three Tests on a Common Ability Scale: Test Equating ... 167

References 171

Index 173

Introduction

When the original book was first published in 1985, the fields of educational measurement and psychometrics were in a transitional period. The majority of practice was based upon the classical test theory developed during the 1920s. However, a new test theory had been developing over the past 40 years that was conceptually more powerful than classical test theory. Based upon items rather than test scores, the new approach was known as item response theory. While the basic concepts of item response theory are straightforward, the underlying mathematics is somewhat advanced compared to that of classical test theory. As a result, it is difficult to examine some of these concepts without performing a large number of calculations to obtain usable information. The original book was designed to provide the reader access to the basic concepts of item response theory freed of the tedious underlying calculations through an Apple II computer program. The second edition of the original published in 2001 used a version of computer program written in Visual Basic 5.0 that could be obtained at http://ericae.net/irt. Readers accustomed to sophisticated statistical and graphics packages might have found it utilitarian, but nevertheless helpful in understanding various facets of the theory. This book now uses R that is a freely available programming language for applied statistics and data visualization. The file folder accompanying the book contains a set of R functions that implement various facets of the theory. These R functions allow the reader to explore the theory at the conceptual level.

The book is organized in a building block fashion. It proceeds from the simple to the complex with each new topic building on the preceding topics. Within each of the eight chapters, a basic concept is presented, the corresponding computer session is explained, and a set of exploratory exercises are defined. Readers are then strongly encouraged to use the computer session to explore the concept through a series of exercises. A final section of each chapter, called "Things to Notice," lists some of the characteristics of the concept that you should have noticed and some of the conclusions that you should have reached. If you do not understand the logic underlying something in this section, you can return to the computer session and try new variations and explorations until clarity is achieved.

When finished with the book and the computer sessions, the reader should have a good working knowledge of the fundamentals of item response theory. This book emphasizes the basics, minimizes the amount of mathematics, and does not pursue technical details that are of interest only to the specialist. In some sense, you will be shown only "what you need to know" rather than all the glorious details of the theory. Upon completion of this book, the reader should be able to interpret test results that have been analyzed under item response theory by means of programs such as WINSTEPS (Linacre 2015), BILOG-MG (Zimowski et al. 2002), and PCLOGIST (Wingersky et al. 1999). Note that WINSTEPS is a current descendant of BICAL (Wright and Mead 1976), BILOG-MG is the extended version of BILOG (Mislevy and Bock 1984), and PCLOGIST is the personal computer version of LOGIST (Wingersky et al. 1982). In order to employ the theory in a practical setting, the reader should study more advanced books on the applications of the theory such as Baker and Kim (2004), de Ayala (2009), Embretson and Reise (2000), Nering and Ostini (2010), Reckase (2009), Thissen and Wainer (2001), and van der Linden and Glas (2000) as well as some earlier books including Hambleton and Swaminathan (1984), Hambleton et al. (1991), Wright and Stone (1979), and Hulin et al. (1983).

Getting Started

R is a software package for data analysis and graphical representation. R provides the language, functions, and the computing environment in one convenient package. The main uniform resource locator (URL; i.e., webpage) of R is

http://www.r-project.org

You can download R by clicking one of the Comprehensive R Archive Network (CRAN) mirror sites in

http://cran.r-project.org/mirrors.html

and following the instruction shown on your computer screen for your own operating system of Linux, Macintosh, or Windows. The base subdirectory contains the binaries for R. Appendix A contains a brief introductory summary of R.

After installing R on your computer, you can perform all activities shown in the respective chapters and the computer sessions by typing in the R command lines exactly shown in the book. Alternatively, for larger R command lines that may contain R functions in the book, you can obtain and use a zipped file folder (BIRTRFunctions.zip) that contains R scripts from the publisher's web site.

Chapter 1
The Item Characteristic Curve

1.1 Introduction

In many educational and psychological measurement situations there is an underlying variable of interest. This variable is often something that is intuitively understood, such as "intelligence." People can be described as being bright or average and the listener has some idea as to what the speaker is conveying about the object of the discussion. Similarly, one can talk about scholastic ability and its attributes such as gets good grades, learns new material easily, relates various sources of information, and uses study time effectively. In academic areas, one can use descriptive terms such as reading ability and arithmetic ability. Each of these is what psychometricians refer to as an unobservable or latent trait. While such a variable is easily described and knowledgeable persons can list its attributes, it cannot be measured directly as can height or weight, since the variable is a concept rather than a physical dimension. A primary goal of educational and psychological measurement is the determination of how much of such a latent trait a person possesses. Since most of the research has dealt with variables such as scholastic, reading, mathematical, and arithmetic abilities, the generic term "ability" is used within item response theory to refer to such latent traits.

If one is going to measure how much of a latent trait a person has, it is necessary to have a scale of measurement, that is, a ruler having a given metric. For a number of technical reasons, defining the scale of measurement, the numbers on the scale, and the amount of the trait that the numbers represent is a very difficult task. For the purposes of the first six chapters, this problem shall be solved by simply defining an arbitrary underlying ability scale. It will be assumed that, whatever the ability, it can be measured on a scale having a midpoint of zero, a unit of measurement of one, and a range from negative infinity to positive infinity. Since there is a unit of measurement and an arbitrary zero point, such a scale is referred to as existing at an interval level of measurement. The underlying idea here is that, if one could physically ascertain the ability of a person, this ruler would be used to tell how much ability a given person has, and the ability of several persons could be compared. While the theoretical range of ability is from negative infinity to positive infinity, practical considerations usually limit the range of values from, say, −3 to +3. Consequently, the discussions in the text and the computer sessions will only deal with ability values within this range. However, you should be aware that values beyond this range are possible.

The usual approach taken to measure an ability is to develop a test consisting of a number of items (i.e., questions). Each of these items measures some facet of the particular ability of interest. From a purely technical point of view such items should be free response items where the examinee can write any response that seems appropriate. The person scoring the test then must decide whether the response is correct or not. When the item response is determined to be correct, the examinee receives a score of one, an incorrect answer receives a score of zero, that is, the item is dichotomously scored. Under classical test theory, the examinee's raw test score would be the sum of the scores received on the items in the test. Under item response theory, the primary interest is in whether an examinee got each individual item correct or not rather than in the raw test score. This is because the basic concepts of item response theory rest upon the individual items of a test rather than upon some aggregate of the item responses such as a test score.

From a practical point of view, free response items are difficult to use in a test. In particular, they are difficult to score in a reliable manner. As a result, most tests used under item response theory consist of multiple-choice items. These are scored dichotomously with the correct answer receiving a score of one and each of the distractors yielding a score of zero. Items scored dichotomously are often referred to as binary items.

1.2 The Item Characteristic Curve

A reasonable assumption is that each examinee responding to a test item possesses some amount of the underlying ability. Thus, one can consider each examinee to have a numerical value, a score, that places the examinee somewhere on the ability scale. This ability score will be denoted by the Greek letter theta, θ. At each ability level there will be a certain probability that an examinee with that ability will give a correct answer to the item. This probability will be denoted by $P(\theta)$. In the case of a typical test item, this probability will be small for examinees of low ability and large for examinees of high ability. If one plotted $P(\theta)$ as a function of ability, the result would be a smooth S-shaped curve such as shown in Fig. 1.1. The probability of correct response is near zero at the lowest levels of ability and increases until at the highest levels of ability the probability of correct response approaches unity. This S-shaped curve describes the relationship between the probability of correct response to an item and the ability scale. In item response theory, it is known as the item characteristic curve. Each item in a test will have its own item characteristic curve.

1.3 Item Difficulty and Item Discrimination

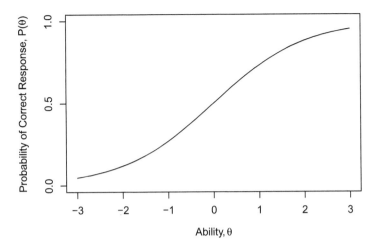

Fig. 1.1 A typical item characteristic curve

The item characteristic curve is the basic building block of item response theory and all the other constructs of the theory depend upon this curve. Because of this, considerable attention will be devoted to this curve and its role within the theory. There are two technical properties of an item characteristic curve that are used to describe it. The first is the difficulty of the item. Under item response theory, the difficulty of an item describes where the item functions along the ability scale. For example, an easy item functions among the low-ability examinees while a hard item would function among the high-ability examinees; thus, item difficulty is a location index. The second technical property is the discrimination of an item, which describes how well an item can differentiate between examinees having abilities below the item location and those having abilities above the item location. This property essentially reflects the steepness of the item characteristic curve in its middle section. The steeper the curve the better the item can discriminate. The flatter the curve the less the item is able to discriminate since the probability of correct response at low ability levels is nearly the same as it is at high ability levels. Using these two descriptors, one can describe the general form of the item characteristic curve. These descriptors are also used to discuss the technical properties of an item. It should be noted that these two properties say nothing about whether the item really measures some facet of the underlying ability or not; that is a question of validity. These two properties simply describe the form of the item characteristic curve.

1.3 Item Difficulty and Item Discrimination

The idea of item difficulty as a location index will be examined first. In Fig. 1.2, three item characteristic curves are presented on the same graph. All have the same level of item discrimination but differ with respect to item difficulty. The left-hand

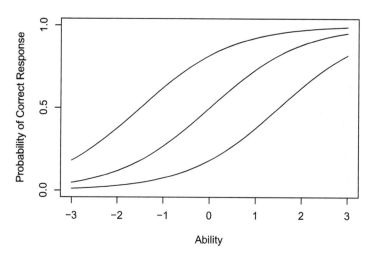

Fig. 1.2 Three item characteristic curves with the same item discrimination but different levels of item difficulty

curve represents an easy item because the probability of correct response is high for low-ability examinees and approaches 1 for high-ability examinees. The center curve represents an item of medium item difficulty because the probability of correct response is low at the lower ability levels, around 0.5 in the middle of the ability scale, and near 1 at the highest ability level. The right-hand curve represents a hard item. The probability of correct response is low for most of the ability scale and increases only when the higher ability levels are reached. Even at the highest ability level shown (i.e., $\theta = 3$) the probability of correct response is only 0.8 for the most difficult item.

The concept of item discrimination is illustrated in Fig. 1.3. This figure contains three item characteristic curves having the same item difficulty but differing with respect to item discrimination. The upper curve on the positive ability range has a high level of item discrimination since the curve is quite steep in the middle where the probability of correct response changes very rapidly as ability increases. Just a short distance to the left of the middle of the curve, the probability of correct response is much less than 0.5; and a short distance to the right, the probability is much greater than 0.5. The middle curve represents an item with a moderate level of item discrimination. The slope of this curve is much less than the previous curve and the probability of correct response changes less dramatically than the previous curve as the ability level increases. However, the probability of correct response is near zero for the lowest-ability examinees and near unity for the highest-ability examinees. The third curve represents an item with low item discrimination. The curve has a very small slope and the probability of correct response changes slowly over the full range of abilities shown. Even at low ability levels, the probability of correct response is reasonably large and it increases only slightly when high ability levels are reached. The reader should be warned that although the figures only show a range of ability from -3 to $+3$, the theoretical range of ability is from negative

1.3 Item Difficulty and Item Discrimination

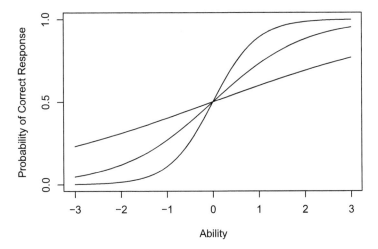

Fig. 1.3 Three item characteristic curves with the same item difficulty but different levels of item discrimination

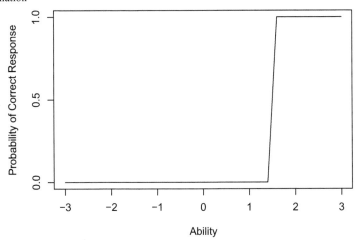

Fig. 1.4 An item that discriminates perfectly at $\theta = 1.5$

infinity to positive infinity. Thus, all item characteristic curves of the type used here actually become asymptotic to a probability of zero at one tail and to unity at the other tail. The restricted range employed in the figures is necessary to fit the curves on the computer screen reasonably and to provide a uniform frame of reference.

One special case is of interest; namely, that of an item with perfect discrimination. The item characteristic curve of such an item is a vertical line at some point along the ability scale. Figure 1.4 shows such an item. To the left of the vertical line at $\theta = 1.5$, the probability of correct response is zero and to the right of the line the probability of correct response is unity. Thus, the item discriminates perfectly between examinees whose abilities are above and below an ability score of 1.5. Such items would be ideal for distinguishing between examinees with abilities just above and below 1.5. However, such an item makes no distinction among those examinees with abilities above 1.5 nor among those examinees with abilities below 1.5.

1.4 Verbal Terms of Item Difficulty and Item Discrimination

At the present point in the presentation of item response theory, the goal is to allow you to develop an intuitive understanding of the item characteristic curve and its properties. In keeping with this goal, item difficulty and item discrimination will be defined in verbal terms. Item difficulty will have the following levels:

- very easy
- easy
- medium
- hard
- very hard

Item discrimination will have the following levels:

- none
- low
- moderate
- high
- perfect

These terms will be used in the computer session to specify item characteristic curves.

1.5 Computer Session

The purpose of this session is to enable you to develop a sense of how the shape of the item characteristic curve is related to item difficulty and item discrimination. To accomplish this, you will be able to select verbal terms describing the item difficulty and item discrimination of an item. The computer program R will then calculate and display the corresponding item characteristic curve on the screen. You should do examples in this section and exercises in the next section, then try various combinations of levels of item difficulty and item discrimination and relate these to the resulting curves. After a bit of such exploratory practice, you should be able to predict what the item characteristic curve will look like for a given combination of item difficulty and item discrimination.

1.5 Computer Session

1.5.1 Procedures for an Example Case

When R is ready for input through the R console window, it prints out its prompt character with an invisible, horizontal space after it:

>

A command line in R will be executed by pressing the enter key:

Such a special character that indicates the end of a command line, usually entered by pressing the enter or return key will be treated as an invisible character here. When an incomplete command line (e.g., the end of the expression cannot have occurred yet) gets the enter key, R prints out the continuation prompt character with an invisible space after it:

+

To improve readability and because a rather long command line can be typed in without pressing the enter key in the middle, a long command line will not be separated by the continuation prompt character but will be continued to the next line with indentation.

The followings are the simplest command lines, each with the R prompt in front, to display an item characteristic curve for an item with medium item difficulty and moderate item discrimination:

```
> theta <- seq(-3, 3, .1)
> bmedium <- 0
> amoderate <- 1
> P <- 1 / (1 + exp(-amoderate * (theta - bmedium)))
> plot(theta, P, type="l")
```

By pressing the enter key in the end of each line, five times as a total, the computer will display an item characteristic curve shown in Fig. 1.5 in the graphics window.

By pressing the enter key in the end of the first line, a sequence of numbers (i.e., a vector) will be created with −3 as a starting number and 3 as an ending number with an increment of 0.1. The length or the total number of elements of the sequence is 61. The name of such a sequence is given as theta, and it is to be done by using the assign function <- for which the less than character < and the hyphen character - cannot be separated with a space. You can read the first line as "theta gets a sequence of numbers . . ." or "a sequence of numbers . . . is saved under the name theta." The first line is equivalent to:

```
> assign("theta", seq(-3,3,.1))
```

or

```
> seq(-3, 3, .1) -> theta
```

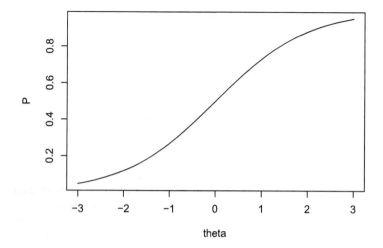

Fig. 1.5 An example item characteristic curve

Use of spaces is optional in the R command line as shown in seq(-3,3,.1). Especially, spaces before and after a comma, a left or right parenthesis, an elementary arithmetic operator (e.g., +, -, *, /), an exponentiation operator (i.e., ∧, **), a common mathematical function (e.g., log, exp, sin, cos, tan, sqrt), a logical operator (e.g., <, <=, >, >=, ==, !=, &, |), and an assign function (e.g., ->, <-) are all optional. To improve readability of the code, at least one space can be put before and after some operators or symbols.

Multiple command lines can be typed in a single line by separating them with a semicolon. Real numbers in the integer form can be entered with a decimal place (i.e., a period) in the end of the number, and the number 0 can be added after the decimal place. The number 0 can be added before the decimal place if the number is less than unity that is to be expressed as a decimal. The second and third lines of the above command lines can be combined as:

> bmedium <- 0; amoderate <- 1

Equivelently, we may use the numbers with a decimal place as:

> bmedium <- 0.0; amoderate <- 1.0

The item characteristic curve is a function of ability and item characteristics, that is, item difficulty and item discrimination. Because ability is a latent variable, it was denoted by θ and the variable name theta was used in the command line. The item difficulty and item discrimination are denoted as numerical values of b and a, respectively, and the verbal terms contain the letters b and a as mnemonic clues. The names of objects in the R command can be constructed with the upper- and lower-case roman letters as an initial character and the digits and the period as any non-initial characters. Names of built-in, intrinsic functions and system variables (e.g., seq, exp, plot, type in the above lines, and c, q, s, t, C, D, F, I, T,

1.5 Computer Session

diff, length, mean, sd, pi, range, rank, time, tree, var, etc.) should not be used as the names of variables or functions of your own. Note that R is case sensitive, so the variable P will be different from the variable p unless both are defined to be equivalent.

The command line

```
P <- 1 / (1 + exp(-amoderate * (theta - bmedium))))
```

will create a vector of 61 values of the probabilities of correct response based on the respective 61 ability points, item difficulty, and item discrimination. We will explore the exact meaning of this function in the subsequent chapter.

The plot based on the 61 sets of points from the ability variable as the horizontal axis, abscissa, and the probability of correct response as a vertical axis, ordinate, can be constructed via:

```
> plot(theta, P)
```

With the enter key pressed, the above line can open an R graphics window that contains the plot of the two variables. The default setting of the function plot will create a plot with a symbol o as each point. Because we want to have a plot with connected lines that ultimately yield a curve, the optional argument type="l" was added in the earlier command line. Whenever R opens up its graphics window, it treats the graphics window as a current window. If you want to continue to use command lines, you may click the R console window (especially the caption on the top of the R console window or any inside portion of the R console window) to make it current before you type in a new command line.

You may notice that the number of ticks used in the default setting of the function plot may not be an appropriate one you want to use in your own figure. The numbers of ticks in the horizontal and vertical axes can be modified with the use of the graphical parameters function par and its labels argument lab, for example:

```
> par(lab=c(7,3,3))
```

The labels argument was defined by the three parameters in the above line. The set of the three parameters (i.e., the three numbers separated with two commas in the combine function c) specifies the number of ticks on the horizontal axes to be 7, that on the vertical axes to be 3, and the length of characters in the labels to be 3 (but the character length will be most likely ignored in R). If you want to explore or read the full description of the R function, for example par, you can obtain it by typing:

```
> ?par
```

or

```
> help(par)
```

Assuming that your computer is connected to Internet, such a command line will open up a file in html, the HyperText Markup Language, that explains the function in a default browser you are using.

It can be noticed in Fig. 1.5 that the variable names are appeared as the respective labeling texts along the horizontal and vertical axes. You can change them with the arguments of, for example, xlab="Ability" and ylab="Probability of Correct Response" in the function plot. The ranges of the horizontal and vertical variables can be precisely controlled with the use of the arguments and parameters of xlim=c(-3,3) and ylim=c(0,1), by specifying the lower limit number and the upper limit number in the combine or concatenate function c.

The following command lines can be used to obtain Fig. 1.1:

```
> par(lab=c(7,3,3))
> theta <- seq(-3, 3, .1)
> b <- 0
> a <- 1
> P <- 1 / (1 + exp(-a * (theta - b)))
> plot(theta, P, type="l", xlim=c(-3,3), ylim=c(0,1),
    xlab=expression(paste("Ability, ",theta)),
    ylab=expression(paste(
    "Probability of Correct Response, P(",theta,")")))
```

As shown in Fig. 1.1, you may use the function expression to add characters in Greek. We may simply use xlab="Ability" and ylab="Probability of Correct Response". You can add the main heading on the top of the plot and the subheading to the bottom of the plot with the arguments in the function plot. These will be helpful when you are making figures for a presentation purpose. For example, you may construct a figure (n.b., a figure is not displayed here) that is more elaborate but nevertheless similar to Fig. 1.5 using the following command lines:

```
> par(lab=c(7,3,3))
> theta <- seq(-3, 3, .1)
> bmedium <- 0
> amoderate <- 1
> P <- 1 / (1 + exp(-amoderate * (theta - bmedium)))
> plot(theta, P, type="l", xlim=c(-3,3), ylim=c(0,1),
    xlab="Ability", ylab="Probability of Correct Response",
    main="Figure 1. An Item Characteristic Curve with
    Medium Item Difficulty and Moderate Item Discrimination",
    sub="See Baker and Kim (2016).")
```

1.5.2 An R Function for Item Characteristic Curves

It is possible to create your own R functions. Each time the plot of the item characteristic curve is created, you may notice that a nearly identical set of R command lines are executed. To avoid the repetition of typing in the same command lines, an R function for plotting an item characteristic curve can be constructed. Consider the following function named `iccplot`:

```
> iccplot <- function(b, a) {
    par(lab=c(7,3,3))
    theta <- seq(-3, 3, .1)
    P <- 1 / (1 + exp(-a * (theta - b)))
    plot(theta, P, type="l", xlim=c(-3,3), ylim=c(0,1),
       xlab="Ability", ylab="Probability of Correct Response")
  }
```

After defining the function, by specifying the numerical values of item difficulty and item discrimination, a plot of the item characteristic curve can be constructed in the R graphics window. For example, you may use the following line to obtain a plot of item characteristic curve with medium item difficulty and moderate item discrimination:

```
> iccplot(0, 1)
```

The two arguments, b and a, are the named actual arguments in R. When the numerical values are specified without the arguments, R recognizes the first number to be the value of item difficulty and the second number to be the value of item discrimination. The arguments in the function can be specified in arbitrary order by exactly defining them with names. You can obtain the same plot by typing:

```
> iccplot(a=1, b=0)
```

Instead of using the numerical values to define item difficulty and item discrimination, the verbal terms described earlier can be used to plot item characteristic curves. The numerical definitions of the verbal terms for item difficulty are as follows:

```
> bveryeasy <- -2.625
> beasy <- -1.5
> bmedium <- 0
> bhard <- 1.5
> bveryhard <- 2.625
```

The numerical definitions of the verbal terms for item discrimination are as follows:

```
> anone <- 0
> alow <- 0.4
> amoderate <- 1
> ahigh <- 2.1
> aperfect <- 999
```

The following command lines now can display two item characteristic curves in the R graphics window (see Fig. 1.6). Note that you should click the R console window after pressing the enter key in the end of the first line, that is, after creating the R graphics window.

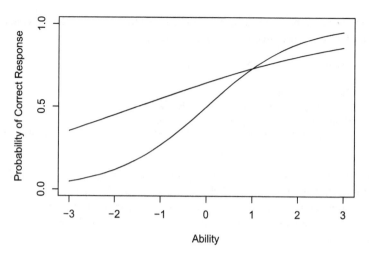

Fig. 1.6 Two item characteristic curves

```
> iccplot(bmedium, amoderate)
> par(new=T)
> iccplot(beasy, alow)
```

The first item characteristic curve is for an item with medium item difficulty and moderate item discrimination. The line

```
> par(new=T)
```

is an overlying figures parameter and equivalent to:

```
> par(new=TRUE)
```

The line resets a graphics parameter new to be TRUE, meaning that the R graphics window is treated now as a new graphics device, so it is assumed that there are currently no plots on it. In such a case, a call to a high-level plotting function could not erase the canvas of the graphics window before putting up a plot. After entering the third line, you may notice that the second item characteristic curve was overlaid on the same graph as the previous curve and you can compare the two. The new curve is rather flat and has higher probabilities of correct response in the lower range of abilities than did the previous item characteristic curve. This is because it was an easier item and low-ability examinees should do well on it. The low item discrimination shows up in the curve having only a slight bend over the range of ability scores employed. At the high ability levels the probability of correct response was somewhat lower than that of the previous item. This is reflection of the lower discrimination on the new item. When the function is specified without par(new=T), a new plot is created in the R graphics window.

Note that you can obtain a figure that contains several item characteristic curves. For example, to obtain Fig. 1.2 you can type the following command lines with clicking the console window after the first line and the third line, respectively:

```
> iccplot(beasy, amoderate)
> par(new=T)
> iccplot(bmedium, amoderate)
> par(new=T)
> iccplot(bhard, amoderate)
```

1.6 Exercises

For the exercises, it is assumed that you have defined the function `iccplot` and the numerical values of the verbal terms of item difficulty and item discrimination by typing in the respective command lines.

1. An item with easy item difficulty and high item discrimination is to be plotted.

    (a) Use the function `iccplot` to plot an item characteristic curve of an item with easy item difficulty and high item discrimination.
    (b) From the graph it can be seen that the probability of correct response will be rather high over most of the ability scale. The item characteristic curve will be steep in the lower part of the ability scale.
    (c) After you have studied the curve, make sure to click the R console window to make it current and to type in a new command line.
    (d) The next graph will be plotted in a new graphics window.

2. An item with hard item difficulty and low item discrimination is to be plotted.

    (a) Use the function `iccplot` to plot an item characteristic curve of an item with hard item difficulty and low item discrimination.
    (b) From the graph it can be seen that the probability of correct response will have a low general level over most of the ability scale. The item characteristic curve will not be very steep.
    (c) After you have studied the curve, make sure to click the R console window to make it current and to type in a new command line.
    (d) The next graph will be plotted in a new graphics window.

3. An item with medium item difficulty and low item discrimination is to be plotted.

    (a) Use the function `iccplot` to plot an item characteristic curve of an item with medium item difficulty and low item discrimination.
    (b) From the graph it can be seen that the probability of correct response will be between 0.2 and 0.8 over the range of ability shown. The item characteristic curve will be nearly linear over the range of ability employed.
    (c) After you have studied the curve, make sure to click the R console window to make it current and to type in a new command line.
    (d) The next graph will be plotted in a new graphics window.

4. In this exercise, all the items will have the same level of item difficulty but different levels of item discrimination. The intent is to relate the steepness of the curve to the level of item discrimination.

   (a) Use the function `iccplot` to plot an item characteristic curve of an item with medium item difficulty and moderate item discrimination.
   (b) From the graph it can be seen that the probability of correct response will be small at low ability levels and large at high ability levels. The item characteristic curve will be moderately steep in the middle part of the ability scale.
   (c) After you have studied the curve, make sure to click the R console window to make it current and to type in a new command line.
   (d) The next graph will be plotted on the same graph in the graphics window. Type in:

   > `par(new=T)`

   (e) Now repeat steps a through d several times using medium item difficulty for each item and item discrimination levels of your choice (e.g., none, low, high, perfect).
   (f) The next graph will be plotted in a new graphics window.

5. In this exercise, all the items will have the same level of item discrimination but different levels of item difficulty. The intent is to relate the location of the item on the ability scale to its level of item difficulty.

   (a) Use the function `iccplot` to plot an item characteristic curve of an item with very easy item difficulty and moderate item discrimination.
   (b) From the graph it can be seen that the probability of correct response will be reasonably large over most of the ability scale. The item characteristic curve will be moderately steep in the lower part of the ability scale.
   (c) After you have studied the curve, make sure to click the R console window to make it current and to type in a new command line.
   (d) The next graph will be plotted on the same graph in the graphics window. Type in:

   > `par(new=T)`

   (e) Now repeat steps a through d several times using items with moderate item discrimination and item difficulty levels of your choice (e.g., easy, medium, hard, very hard).
   (f) The next graph will be plotted in a new graphics window.

6. Experiment with various combinations of item difficulty and item discrimination of your own choice until you are confident that you can predict the shape of the item characteristic curve corresponding to the levels chosen. You may find it useful to make a rough sketch of what you think the curve will look like before you have the computer display it on the screen.

1.7 Things to Notice

1. When item discrimination is less than moderate, the item characteristic curve is nearly linear and appears rather flat.
2. When item discrimination is greater than moderate, the item characteristic curve is S-shaped and rather steep in its middle section.
3. When item difficulty is less than medium, most of the item characteristic curve has a probability of correct response that is greater than 0.5.
4. When item difficulty is greater than medium, most of the item characteristic curve has a probability of correct response less than 0.5.
5. Regardless of the level of item discrimination, item difficulty locates the item along the ability scale. Therefore item difficulty and item discrimination are independent of each other.
6. When an item has no item discrimination, all choices of item difficulty yield the same horizontal line at a value of $P(\theta) = 0.5$. This is because the value of item difficulty for an item with no item discrimination is undefined.
7. If you have been very observant, you may have noticed the point at which $P(\theta) = 0.5$ corresponds to item difficulty. When an item is easy, this value occurs at a low ability level. When an item is hard, this value corresponds to a high ability level.

Chapter 2
Item Characteristic Curve Models

2.1 Introduction

In the first chapter the properties of the item characteristic curve were defined in terms of verbal descriptors. While this is useful to obtain an intuitive understanding of item characteristic curves, it lacks the precision and rigor needed by a theory. Consequently, in this chapter the reader will be introduced to three mathematical models for the item characteristic curve. These models provide mathematical equations for the relation of the probability of correct response to ability. Each model employs one or more item parameters whose numerical values define a particular item characteristic curve. Such mathematical models are needed if one is to develop a measurement theory that can be rigorously defined and is amenable to further growth. In addition, these models and their parameters provide a vehicle for communicating information about an item's technical properties. For each of the three models, the mathematical equation will be used to compute the probability of correct response at several ability levels. Then the graph of the corresponding item characteristic curve will be shown. The goal of the chapter is to have you develop a sense of how the numerical values of the item parameters for a given model relate to the shape of the item characteristic curve.

2.2 The Two-Parameter Model

Under item response theory the standard mathematical model for the item characteristic curve is the cumulative form of the logistic function. It defines a family of curves having the general shape of the item characteristic curves shown in the first chapter. The logistic function was first derived in 1844 by Pierre François Verhulst and has been widely used in the biological sciences to model the growth of plants and animals from birth to maturity. It was first used as a model for the item characteristic curve in the late 1950s and, because of its simplicity, has become the preferred model. The equation for the two-parameter logistic model is given in Eq. (2.1) below:

$$P(\theta) = \frac{1}{1+e^{-L}} = \frac{1}{1+e^{-a(\theta-b)}}, \qquad (2.1)$$

where

e is the base of the natural logarithm that is a constant 2.718,
b is the item difficulty parameter,
a is the item discrimination parameter,[1]
$L = a(\theta - b)$ is the logistic deviate (logit), and
θ is an ability level.

This equation represents a family of curves whose individual members are defined by specific numerical values of the item parameters b and a; hence, it is called the two-parameter model. It is the model that was used in Chap. 1.

The item difficulty parameter, denoted by b, is defined as the point on the ability scale at which the probability of correct response to the item is 0.5. The theoretical range of the values of this parameter is $-\infty \leq b \leq +\infty$ (i.e., a set of extended real numbers). However, typical values have the range of $-3 \leq b \leq +3$.

Due to the shape of the item characteristic curve, the slope of the curve changes as a function of the ability level and reaches a maximum value when the ability level equals the item difficulty parameter. Because of this, the item discrimination parameter does not represent the general slope of the item characteristic curve as was indicated in Chap. 1. The technical definition of the item discrimination parameter is beyond the level of this book. However, a usable definition is that this parameter is proportional to the slope of the item characteristic curve at $\theta = b$. The actual slope at $\theta = b$ is $a/4$ under the two-parameter model, but considering a to be the slope at b is an acceptable approximation that makes interpretation of the item discrimination parameter easier in practice. The theoretical range of the values of this parameter is $-\infty \leq a \leq +\infty$, but the usual range seen in practice is -2.80 to $+2.80$.

---

[1] In much of the item response theory literature, the logistic value of the item discrimination parameter a is divided by 1.702 or 1.7 to obtain the corresponding normal ogive model value. This is done to make the two-parameter logistic ogive similar to the normal ogive. However, this was not done in this book as it introduces two frames of reference for interpreting the numerical values of the item discrimination parameter. All item parameters in this book and the associated computer programs are interpreted in terms of the logistic function.

2.2 The Two-Parameter Model

2.2.1 Computational Example

To illustrate how the two-parameter model is used to compute the points on an item characteristic curve, consider the following example problem. The values of the item parameters are:

$b = 1.0$ is the item difficulty parameter
$a = 0.5$ is the item discrimination parameter

The illustrative computation is performed at the ability level:

$\theta = -3.0$

The first term to be computed is the logistic deviate (logit), L, where:

$L = a(\theta - b)$

Substituting the appropriate values yields:

$L = 0.5(-3.0 - 1.0) = -2.0$

The next term computed is e raised to the power $-L$. If you have a pocket calculator that can compute e^x or $\exp(x)$ you can verify this calculation. Substituting yields:

$e^{-L} \equiv \exp(-L) = \exp(2.0) = 7.389$

Now the denominator of Eq. (2.1) can be computed as:

$1 + \exp(-L) = 1 + 7.389 = 8.389$

Finally, the value of $P(\theta)$ is:

$P(\theta) = 1/[1 + \exp(-L)] = 1/8.389 = 0.119$

Thus, at an ability level of $\theta = -3.0$, the probability of responding correctly to this item is 0.119.

From the above, it can be seen that computing the probability of correct response at a given ability level is very easy using the logistic model. Table 2.1 shows the calculations for this item at seven ability levels evenly spaced over the range of ability levels from -3 to $+3$. You should perform the computations at several of these ability levels to become familiar with the procedure.

Table 2.1 Item characteristic curve calculations under the two-parameter model, $b = 1.0$ and $a = 0.5$

Ability, θ	Logit, L	$\exp(-L)$	$1 + \exp(-L)$	$P(\theta)$
−3.0	−2.0	7.389	8.389	0.119
−2.0	−1.5	4.482	5.482	0.182
−1.0	−1.0	2.718	3.718	0.269
0.0	−0.5	1.649	2.649	0.378
1.0	0.0	1.000	2.000	0.500
2.0	0.5	0.607	1.607	0.622
3.0	1.0	0.368	1.368	0.731

Fig. 2.1 The item characteristic curve for the two-parameter model with $b = 1.0$ and $a = 0.5$

The item characteristic curve for the item of Table 2.1 is shown in Fig. 2.1. The vertical dotted line corresponds to the value of the item difficulty parameter.

2.3 The Rasch Model

The next model of interest was first published by the Danish mathematician Georg Rasch in the 1960s. Rasch approached the analysis of test data from a probability theory point of view. Although he started from a very different frame of reference, the resultant item characteristic curve model was a logistic model. In Chap. 8, Rasch's approach will be explored in greater detail; our present interest is only in his item characteristic curve model. Under this model, the item discrimination parameter of the two-parameter model is fixed at a value of $a = 1$ for all items and only the item difficulty parameter can take on different values. Because of this, the Rasch model is often referred to as the one-parameter logistic model.

The equation for the Rasch model is given by the following:

$$P(\theta) = \frac{1}{1 + e^{-1(\theta - b)}}, \tag{2.2}$$

where

b is the item difficulty parameter and
θ is the ability level.

It should be noted that the item discrimination parameter was used in Eq. (2.2). But because it always has a value of 1 it usually is not shown in the formula.

2.3 The Rasch Model

2.3.1 Computational Example

Again the illustrative computations for the model will be done for the single ability level:

$$\theta = -3.0$$

The value of the item difficulty parameter is:

$$b = 1.0$$

The first term computed is the logit, L, where:

$$L = a(\theta - b) = \theta - b$$

Substituting the appropriate values yields:

$$L = -3.0 - 1.0 = -4.0$$

Next, the e raised to the $-L$ term is computed, giving:

$$\exp(-L) = 54.598$$

The denominator of Eq. (2.2) can be computed as:

$$1 + \exp(-L) = 1 + 54.598 = 55.598$$

Finally, the value of $P(\theta)$ can be obtained and is:

$$P(\theta) = 1/[1 + \exp(-L)] = 1/55.598 = 0.018$$

Thus, at an ability level of $\theta = -3.0$, the probability of responding correctly to this item is 0.018.

Table 2.2 shows the calculations at seven ability levels. You should perform the computations at several other ability levels to become familiar with the model and the procedure.

The item characteristic curve for the item of Table 2.2 is shown in Fig. 2.2. The vertical dotted line corresponds to the value of the item difficulty parameter.

Table 2.2 Item characteristic curve calculations under the Rasch model, $b = 1.0$

Ability, θ	Logit, L	$\exp(-L)$	$1 + \exp(-L)$	$P(\theta)$
−3.0	−4.0	54.598	55.598	0.018
−2.0	−3.0	20.086	21.086	0.047
−1.0	−2.0	7.389	8.389	0.119
0.0	−1.0	2.718	3.718	0.269
1.0	0.0	1.000	2.000	0.500
2.0	1.0	0.368	1.368	0.731
3.0	2.0	0.135	1.135	0.881

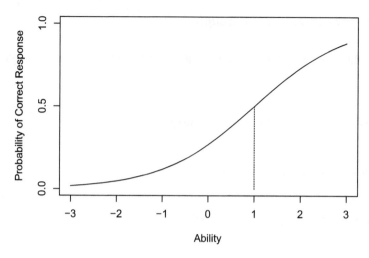

Fig. 2.2 The item characteristic curve for the Rasch model with $b = 1.0$

2.4 The Three-Parameter Model

One of the facts of life in testing is that examinees will get items correct by guessing. Thus, the probability of correct response includes a small component that is due to guessing. Neither of the two previous item characteristic curve models took the guessing phenomenon into consideration. Birnbaum (1968) modified the two-parameter logistic model to include a parameter that represents the contribution of guessing to the probability of correct response. Unfortunately, in so doing, some of the nice mathematical properties of the logistic function were lost. Nevertheless the resulting model has become known as the three-parameter logistic model even though it technically is no longer a logistic model. The equation for the three-parameter model is:

$$P(\theta) = c + (1-c)\frac{1}{1+e^{-a(\theta-b)}}, \tag{2.3}$$

where

b is the item difficulty parameter,
a is the item discrimination parameter,
c is the guessing parameter, and
θ is the ability level.

The parameter c is the probability of getting the item correct by guessing alone. It is important to note that by definition the value of c does not vary as a function of the ability level. Thus, the lowest and highest ability examinees may have the same probability of getting the item correct by guessing. The parameter c has a theoretical range of $0 \leq c \leq 1.0$, but in practice values above 0.35 are not considered acceptable, hence the range $0 < c < 0.35$ is used here.

2.4 The Three-Parameter Model

A side effect of using the guessing parameter c is that the definition of the item difficulty parameter is changed. Under the previous two models, b was the point on the ability scale at which the probability of correct response was 0.5. But now the lower limit of the item characteristic curve is the value of c rather than zero. The result is that the item difficulty parameter is the point on the ability scale where

$$P(\theta) = c + (1 - c)(0.5) = (1 + c)/2.$$

This probability is halfway between the value of c and 1.0. What has happened here is that the parameter c has defined a floor to the lowest value of the probability of correct response. Thus, the item difficulty parameter defines the point on the ability scale where the probability of correct response is halfway between this floor and 1.0.

The item discrimination parameter a can still be interpreted as being proportional to the slope of the item characteristic curve at the point $\theta = b$. However, under the three-parameter model, the slope of the item characteristic curve at $\theta = b$ is actually $a(1-c)/4$.

While these changes in the definitions of item parameters b and a seem slight, they are important when interpreting the results of test analyses.

2.4.1 Computational Example

The probability of correct response to an item under the three-parameter model will be shown for the following item parameter values:

$b = 1.5$
$a = 1.3$
$c = 0.2$

The ability level is:

$\theta = -3.0$

The logit is:

$L = a(\theta - b) = 1.3(-3.0 - 1.5) = -5.85$

Next, the e raised to the $-L$ term is:

$\exp(-L) = 347.234$

The denominator of the term in the right-hand side of Eq. (2.3) is:

$1 + \exp(-L) = 1 + 347.234 = 348.234$

Then the right-hand side term yields:

$1/[1 + \exp(-L)] = 1/348.234 = 0.0029$

Up to this point the computations are exactly the same as those for the two-parameter model with $b = 1.5$ and $a = 1.3$. But now the guessing parameter enters the picture. From Eq. (2.3) we have:

$$P(\theta) = c + (1 - c)(0.0029)$$

With $c = 0.2$ the value of $P(\theta)$ is:

$$P(\theta) = 0.2 + (1 - 0.2)(0.0029) = 0.2 + (0.8)(0.0029) = 0.2 + (0.0023) = 0.2023$$

Thus, at an ability level of $\theta = -3.0$, the probability of responding correctly to this item is 0.2023.

Table 2.3 shows the calculations at seven ability levels. Again, you are urged to perform the above calculations at several other ability levels to become familiar with the model and the procedure.

The item characteristic curve for the item of Table 2.3 is shown in Fig. 2.3. The vertical dotted line corresponds to the value of the item difficulty parameter.

Table 2.3 Item characteristic curve calculations under the three-parameter model, $b = 1.5$, $a = 1.3$, and $c = 0.2$

Ability, θ	Logit, L	$\exp(-L)$	$1 + \exp(-L)$	$P(\theta)$
-3.0	-5.85	347.234	348.234	0.202
-2.0	-4.55	94.632	95.632	0.208
-1.0	-3.25	25.790	26.790	0.230
0.0	-1.95	7.029	8.029	0.300
1.0	-0.65	1.916	2.916	0.474
2.0	0.65	0.522	1.522	0.726
3.0	1.95	0.142	1.142	0.900

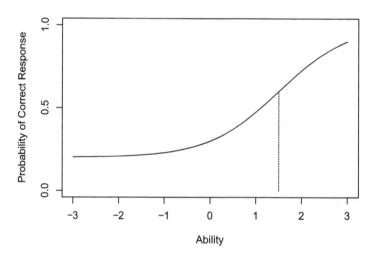

Fig. 2.3 The item characteristic curve for the three-parameter model with $b = 1.5$, $a = 1.3$, and $c = 0.2$

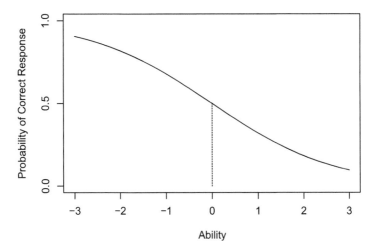

Fig. 2.4 An item with negative discrimination under the two-parameter model with $b = 0$ and $a = -0.75$

2.5 Negative Discrimination

While most test items will discriminate in a positive manner; that is, the probability of correct response increases as the ability level increases, some items have negative discrimination. In such items, the probability of correct response decreases as the ability level increases from low to high. Figure 2.4 depicts such an item.

Items with negative discrimination occur in two ways. First, the incorrect response to a two-choice item will always have a negative item discrimination parameter if the correct response has a positive value. Second, sometimes the correct response to an item will yield a negative item discrimination parameter. This tells you that something is wrong with the item. Either it is poorly written or there is some misinformation prevalent among the high-ability students. In any case, it is a warning that the item needs some attention. For most of the item response theory topics of interest, the value of the item discrimination parameter will be positive. Figure 2.5 shows the item characteristic curves for the correct and incorrect responses to a binary item.

It should be noted that the two item characteristic curves have the same value for the item difficulty parameter ($b = 1.0$) and the item discrimination parameters have the same absolute value. However, the item discrimination parameters have opposite signs, with the correct response being positive and the incorrect response being negative.

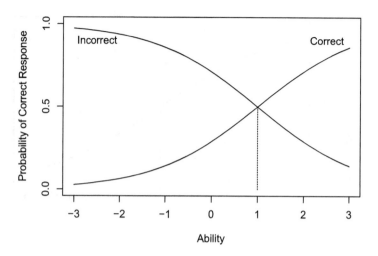

Fig. 2.5 Item characteristic curves for the correct ($b = 1.0$, $a = 0.9$) and incorrect responses ($b = 1.0$, $a = -0.9$) to a binary item

Table 2.4 Labels for item discrimination parameter values

Verbal label	Range of values	Typical value
None	0	0.00
Very low	0.01–0.34	0.18
Low	0.35–0.64	0.50
Moderate	0.65–1.34	1.00
High	1.35–1.69	1.50
Very high	>1.70	2.00
Perfect	$+\infty$	$+\infty$

2.6 Guidelines for Interpreting Item Parameter Values

In Chap. 1, verbal labels were used to describe the technical properties of an item characteristic curve. Now the curves can be described via parameters whose numerical values have intrinsic meaning. However, one needs some means of interpreting the numerical values of the item parameters and conveying this interpretation to a non-technical audience. Table 2.4 presents the verbal labels used to describe the item's discrimination can be related to ranges of values of the item discrimination parameter.

These relations hold when one interprets the values of the item discrimination parameter under the logistic model for the item characteristic curve. If the reader wants to interpret the item discrimination parameter under the normal ogive model, divide these values by 1.7.

Establishing an equivalent table for the values of the item difficulty parameter poses some problems. The terms, easy and hard used in Chap. 1, are relative terms that depend upon some frame of reference. As discussed above, the drawback of item difficulty, as defined under classical test theory, was that it was defined relative to a group of examinees. Thus, the same item could be easy for one group and hard for another group. Under item response theory, an item's difficulty is a point on the ability scale where the probability of correct response is 0.5 for one- and two-parameter models and $(1+c)/2$ for the three-parameter model. Because of this, the verbal labels used in Chap. 1 have meaning only with respect to the midpoint of the ability scale.

The proper way to interpret a numerical value of the item difficulty parameter is in terms of where the item functions on the ability scale. The item discrimination parameter can be used to add meaning to this interpretation. The slope of the item characteristic curve is at a maximum at an ability level corresponding to the item difficulty parameter. Thus, the item is doing its best in distinguishing between examinees in the neighborhood of this ability level. Because of this, one can speak of the item functioning at this ability level. For example, an item whose difficulty is -1 functions among the lower-ability examinees. A value of $+1$ denotes an item that functions among higher-ability examinees. Again, the underlying concept is that the item difficulty is a location parameter.

Under the three-parameter model, the numerical value of the guessing parameter c is interpreted directly since it is a probability. For example, $c = 0.12$ simply means that at all ability levels the probability of getting the item correct by guessing alone is 0.12.

2.7 Computer Session

The purpose of this session is to enable you to develop a sense of the dependence of the shape of the item characteristic curve upon the model and the numerical values of its parameters. You will be able to set the values of the item parameters under each of the three models and the corresponding item characteristic curve will be shown on the screen via the computer program R. Choosing these values becomes a function of what kind of an item characteristic curve one is trying to define. Conversely, given a set of numerical values of the item parameters for an item characteristic curve, such as provided by a test analysis, you should be able to visualize the form of the item characteristic curve. Such visualization is necessary to properly interpret the technical properties of the item. After doing the exercises and a bit of exploration, you should be able to visualize the form of the item characteristic curve for any of the three models given a set of item parameter values.

Table 2.5 Item characteristic curve calculations under the two-parameter model, $b = -1.0$ and $a = 1.7$

Ability, θ	Logit, L	exp(−L)	1 + exp(−L)	$P(\theta)$
−3.0	−3.4	29.964	30.964	0.032
−2.0	−1.7	5.474	6.474	0.154
−1.0	0.0	1.000	2.000	0.500
0.0	1.7	0.183	1.183	0.846
1.0	3.4	0.033	1.033	0.968
2.0	5.1	0.006	1.006	0.994
3.0	6.8	0.001	1.001	0.999

2.7.1 Procedures for an Example Case

The followings are the R command lines to obtain and display the values of various intermediate terms and the probability of correct response under the two-parameter model with $b = -1.0$ and $a = 1.7$:

```
> theta <- seq(-3, 3, 1)
> b <- -1.0
> a <- 1.7
> L <- a * (theta - b)
> P <- 1 / (1 + exp(-L))
> theta; L; exp(-L); 1 + exp(-L); P
```

By pressing the enter key in the end of each line, the computer will display the respective sets of values in the R console window. The values are reported in Table 2.5.

By pressing the enter key in the end of the first line, a sequence of numbers (i.e., a vector) will be created with −3 as a starting number and 3 as an ending number with an increment of 1. The length or the total number of elements of the sequence is seven. The name of the sequence is assigned as theta. The second and third lines define the value of item difficulty parameter to be −1.0 and the value of item discrimination parameter to be 1.7, respectively. The fourth line defines the logit, L, which is in fact a vector of size seven. A vector P of points on the item characteristic curve under the two-parameter model is calculated in the fifth line. Five command lines are combined with four semicolons in the last line. With the enter key pressed in the end of the last line, the five sets of values are obtained.

Instead of using the function exp directly, we may create two variables expnl (i.e., the exponential function of the negative logit) and opexpnl (i.e., one plus the exponential function of the negative logit) and display the same sets of values:

```
> theta <- seq(-3, 3, 1)
> b <- -1.0
> a <- 1.7
> L <- a * (theta - b)
> expnl <- exp(-L)
> opexpnl <- 1 + expnl
```

2.7 Computer Session

```
> P <- 1 / opexpnl
> theta; L; expnl; opexpnl; P
```

The R function data.frame can be used to combine the vectors into a single data frame, and it allows to nicely display the whole values in an organized fashion. The last command line, assuming that the variables have been created as in the above example, can be replaced with:

```
> data.frame(theta, L, expnl, opexpnl, P)
```

Note that the data frame can be named and displayed by typing its name; for example:

```
> table2p5 <- data.frame(theta, L, expnl, opexpnl, P)
> table2p5
```

The computer will display the table of computations. Study the table for a few minutes to see the relation between the probability of correct response and the ability scores.

The item characteristic curve for the item of Table 2.5 can be obtained from the following command lines (see Fig. 2.6):

```
> par(lab=c(7,3,3))
  theta <- seq(-3, 3, .1)
  b <- -1.0
  a <- 1.7
  P <- 1 / (1 + exp(-a * (theta - b)))
  plot(theta, P, type="l", xlim=c(-3,3), ylim=c(0,1),
    xlab="Ability", ylab="Probability of Correct Response")
  thetai <- b
  pthetai <- 1 / (1 + exp(-a * (thetai - b)))
```

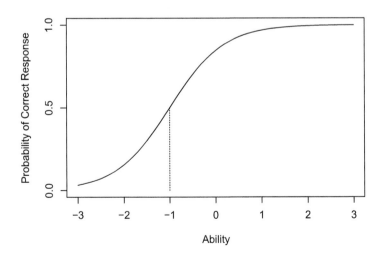

Fig. 2.6 The item characteristic curve for the two-parameter model with $b = -1.0$ and $a = 1.7$

```
vliney <- seq(0, pthetai, .01)
vlinex <- b + vliney * 0
lines(vlinex, vliney, lty=2)
```

The last five lines of the above command lines are used to create the vertical dotted line that indicates the location of the item difficulty parameter. First, the point on the ability scale is defined that is the same as the item difficulty, $\theta_i = b$. The height or ordinate of the item characteristic curve for θ_i is obtained as $P(\theta_i)$. The vertical line is constructed as the collection of the points between two points on the graph, that is, from $(b, 0)$ to $(b, P(\theta_i))$. That vector that contains values in the y axis, vliney, is created with the function seq. The vector that contains the corresponding values in the x axis, vlinex, is created next. This vector contains a set of constant value b, but its size is the same as that of vliney. The function lines construct the line on the existing graph in the graphics window from the collection of two points and depicted it as a dotted line by its optional argument lty=2.

This item functions at an ability level of -1 and the curve is quite steep at that ability level. Notice that the curve is nearly flat above an ability of, say, 1.0. In technical terms it is becoming asymptotic to a value of $P(\theta) = 1.0$. At this juncture the example case is completed. Note that the graphics window is likely to be the current window. Hence, if you want to continue to type in more command lines, make sure that you have clicked the R console window.

2.7.2 An R Function for Item Characteristic Curve Calculations

It is possible to create an R function for displaying item characteristic curve calculations. Consider the following function named icccal:

```
> icccal <- function(b, a, c) {
    if (missing(c)) c <- 0
    if (missing(a)) a <- 1
    theta <- seq(-3, 3, 1)
    L <- a * (theta - b)
    expnl <- exp(-L)
    opexpnl <- 1 + expnl
    P <- c + (1 - c) / opexpnl
    data.frame(theta, L, expnl, opexpnl, P)
  }
```

It is also possible to create an R function for plotting an item characteristic curve. Consider the following function named icc:

```
> icc <- function(b, a, c) {
    if (missing(c)) c <- 0
    if (missing(a)) a <- 1
    par(lab=c(7,3,3))
```

2.8 Exercises

```
        theta <- seq(-3, 3, .1)
        P <- c + (1 - c) / (1 + exp(-a * (theta - b)))
        plot(theta, P, type="l", xlim=c(-3,3), ylim=c(0,1),
          xlab="Ability", ylab="Probability of Correct Response")
        thetai <- b
        pthetai <- c + (1 - c) / (1 + exp(-a * (thetai - b)))
        vliney <- seq(0, pthetai, .01)
        vlinex <- b + vliney * 0
        lines(vlinex, vliney, lty=2)
    }
```

Note that the functions, icccal and icc, are general ones. Both can be used for the Rasch model as well as the two- and three-parameter models.

After typing in these two functions in the R console window, the example case for the table of computations and the graph can be constructed by the following two lines:

```
> icccal(b=-1.0, a=1.7)
> icc(b=-1.0, a=1.7)
```

The two command lines are equivalent to:

```
> icccal(-1.0, 1.7)
> icc(-1.0, 1.7)
```

and:

```
> icccal(a=1.7, b=-1.0)
> icc(a=1.7, b=-1.0)
```

2.8 Exercises

For the exercises, it is assumed that you have defined two functions icccal and icc by typing them in the R console window.

1. This exercise uses the Rasch model to illustrate how the item difficulty parameter locates an item along the ability scale.

   (a) Set the value of the item difficulty parameter to $b = -2.0$. Type in the following command line:

   ```
   > icccal(b=-2.0)
   ```

   (b) The computer will display the table of computations. Study the table for a few minutes to see the relation between the probability of correct response and the ability scores.

   (c) The item characteristic curve will be displayed on the screen by typing in the following command line:

   ```
   > icc(b=-2.0)
   ```

(d) This item will function at an ability level of −2.0 and the curve will be moderately steep at that ability level. Study the plot.
(e) Next, we want to put another item characteristic curve on the same graph. Type in the following command line:

```
> par(new=T)
```

(f) Now set the item difficulty parameter to a value of $b = 0.0$ and repeat steps a through c.
(g) This will place a second curve on the graph.
(h) Now repeat steps e through f using the value of $b = 2.0$ for the item difficulty parameter.
(i) Now there will be three item characteristic curves on the graph. The three dotted lines indicate the values of $P(\theta) = 0.5$ of these curves at the ability levels defined by their values of item difficulty parameters. In the present example the values of the item difficulty parameters are evenly spaced along the ability scale.
(j) At this juncture Exercise 1 is completed.

2. This exercise uses the two-parameter model to illustrate the joint effect of item difficulty and item discrimination upon the shape of item characteristic curve.

   (a) Set the values of the item difficulty parameter to $b = -2.0$ and the item discrimination parameter to 1.0. Type in the following command line:

   ```
   > icccal(b=-2.0, a=1.0)
   ```

   (b) The computer will display the table of computations. Study the table for a few minutes to see the relation between the probability of correct response and the ability scores.
   (c) The item characteristic curve will be displayed on the screen by typing in the following command line:

   ```
   > icc(b=-2.0, a=1.0)
   ```

   (d) The item characteristic curve is located in the low-ability end of the scale and it is moderately steep.
   (e) Next, we want to put another item characteristic curve on the same graph. Type in the following command line:

   ```
   > par(new=T)
   ```

   (f) Now set the item difficulty parameter to a value of $b = 0.0$ and the item discrimination parameter to a value of $a = 1.5$. Then repeat steps a through c.
   (g) This will place a second curve on the graph.
   (h) Now repeat steps e through f using the value of $b = 2.0$ for the item difficulty parameter and the value $a = 0.5$ for the item discrimination parameter.

2.8 Exercises

(i) You should now have three item characteristic curves displayed on the same graph. It should be clear that the value of b locates the item on the ability scale and that a defines the slope. However, in the present example the curves cross because the values of a are different for each item.

(j) At this juncture Exercise 2 is completed.

3. This exercise illustrates the joint effect of the parameter values under the three-parameter model.

   (a) Set the values of the item parameters to $a = 1.0$, $b = -2.0$, and $c = 0.10$. Type in the following command line:

   ```
   > icccal(b=-2.0, a=1.0, c=.10)
   ```

   (b) The computer will display the table of computations. Study the table for a few minutes to see the relation between the probability of correct response and the ability scores.

   (c) The item characteristic curve will be displayed on the screen by typing in the following command line:

   ```
   > icc(b=-2.0, a=1.0, c=.10)
   ```

   (d) The item characteristic curve is located in the low-ability end of the scale and it is moderately steep.

   (e) Next, we want to put another item characteristic curve on the same graph. Type in the following command line:

   ```
   > par(new=T)
   ```

   (f) Now set the values of the item parameters to $b = 0.0$, $a = 1.5$, and $c = 0.20$. Then repeat steps a through c.

   (g) This will place a second curve on the graph.

   (h) Now repeat steps e through f using the values of $b = 2.0$, $a = 0.5$, and $c = 0.30$.

   (i) At this point you should have three item characteristic curves displayed on the graph. Again the values of b locate the items along the ability scale. But the ability level at which $P(\theta) = 0.5$ does not correspond to the value of b but is slightly lower. Recall that under the three-parameter model, b is the point on the ability scale where the probability of correct response is $(1 + c)/2$ rather than 0.5. The slopes of the curves at b reflect the values of a. The lower tails of the three curves approach their values of c at the lowest levels of ability. However, this is not apparent for the curve with $b = -2.0$ as the values of $P(\theta)$ are still rather large at $\theta = -3.0$.

   (j) At this juncture Exercise 3 is completed.

4. The followings are for the additional exercises.

   (a) For each model:

      (i) Select a set of parameter values, and obtain the calculation table.

(ii) Predict what the shape of the item characteristic curve will look like. It can be helpful to make a sketch of the item characteristic curve before the computer shows it on the screen.
(iii) Obtain the display of the curve (it may help to overlay a few curves to get a feeling for the relative effects of changing parameter values).

(b) Repeat this process until you know what kind of item characteristic curve will result from a set of numerical values of the item parameters under each of the models.

2.9 Things to Notice

1. Under the Rasch[2] model, the slope is always the same; only the location of the item changes.
2. Under the two- and three-parameter models, the value of a must become quite large (e.g., >1.7) before the curve is very steep.
3. Under the Rasch and two-parameter models, a large positive value of b results in a lower tail of the curve that approaches zero. But under the three-parameter model, the lower tail approaches the value of c.
4. Under the three-parameter model, the value of c is not apparent when $b < 0$ and $a < 1.0$. However, if a wider range of values of ability were used, the lower tail would approach the value of c.
5. Under all models, curves with a negative value of a are the mirror image of curves with the same values of the remaining parameters and a positive value of a.
6. When $b = -3.0$, only the upper half of the item characteristic curve appears on the graph. When $b = +3.0$, only the lower half of the curve appears on the graph.
7. The slope of the item characteristic curve is the steepest at the ability level corresponding to the item difficulty parameter. Thus, the item difficulty parameter b locates the point on the ability scale where the item functions best.
8. Under the Rasch and two-parameter models, the item difficulty defines the point on the ability scale where the probability of correct response for persons of that ability is 0.5. Under the three-parameter model, the item difficulty parameter defines the point on the ability scale where the probability of correct response is halfway between the value of the parameter c and 1.0. Only when $c = 0$ are these two definitions equivalent.

---

[2]Originally, the Rasch model was referred to as the one-parameter logistic model as the only item parameter was the item difficulty parameter. In recent years, however, a model in which all items share a common value of the item discrimination parameter has been also called the one-parameter logistic model. To avoid confusion, the label Rasch model will be used in this book.

Chapter 3
Estimating Item Parameters

3.1 Introduction

Because the actual values of the parameters of the items in a test are unknown, one of the tasks performed when a test is analyzed under item response theory is to estimate these parameters. The obtained item parameter estimates then provide information as to the technical properties of the test items. To keep matters simple in the following presentation, the parameters of a single item will be estimated under the assumption that the examinees ability scores are known. In reality, these scores are not known, but it is easier to explain how item parameter estimation is accomplished if this assumption is made.

3.2 Maximum Likelihood Estimation of Item Parameters

In the case of a typical test, a sample of N examinees responds to the J items in the test. The ability scores of these examinees will be distributed over a range of ability levels on the ability scale. For present purposes, these examinees will be divided into, say, G groups along the scale so that all the examinees within a given group have the same ability level θ_g and there will be f_g examinees within group g, where $g = 1, 2, \ldots, G$. Within a particular ability score group, r_g examinees answer the given item correctly. Thus, at an ability level of θ_g the observed proportion of correct response is $p(\theta_g) = r_g/f_g$, which is an estimate of the probability of correct response at that ability level. Now the value of r_g can be obtained and $p(\theta_g)$ computed for each of the g ability levels established along the ability scale. If the observed proportions of correct response in each ability group are plotted, the result will be something like that shown in Fig. 3.1.

© Springer International Publishing AG 2017
F.B. Baker, S.-H. Kim, *The Basics of Item Response Theory Using R*,
Statistics for Social and Behavioral Sciences, DOI 10.1007/978-3-319-54205-8_3

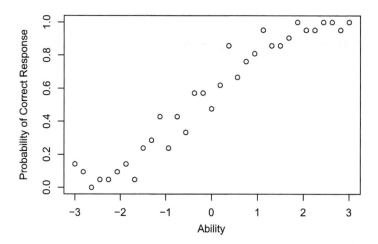

Fig. 3.1 Observed proportions of correct response as a function of ability

The basic task now is to find the item characteristic curve that best fits the observed proportions of correct response. To do so, one must first select a model for the curve to be fitted. Although any of the three logistic models could be used, the two-parameter model will be employed here. The procedure used to fit the curve is based upon maximum likelihood estimation.

Under the maximum likelihood estimation (MLE) procedure, initial values for the item parameters, such as $b = 0.0$ and $a = 1.0$, are established a priori. Then, using these estimates the value of $P(\theta_g)$ is computed at each ability level via the equation for the item characteristic curve model. The agreement of the observed value of $p(\theta_g)$ and the computed value of $P(\theta_g)$ is determined across all ability groups. Then, adjustments to the estimated item parameters are found that result in better agreement between the item characteristic curve defined by the estimated values of the parameters and the observed proportions of correct response. This process of adjusting the estimates is continued until the adjustments get so small that little improvement in the agreement is possible. At this point, the estimation procedure is terminated and the current values of b and a are the item parameter estimates. Given these values, the equation for the item characteristic curve is used to compute the probability of correct response $P(\theta_g)$ at each ability level and the item characteristic curve can be plotted. The resulting curve is the item characteristic curve that best fits the response data for that item. Figure 3.2 shows an item characteristic curve fitted to the observed proportions of correct response shown in Fig. 3.1. The estimated values of the item parameters were $b = -0.39$ and $a = 1.27$ (see Appendix B).

An important consideration within item response theory is whether a particular item characteristic curve model fits the item response data for an item. The agreement of the observed proportions of correct response and those yielded by the

3.2 Maximum Likelihood Estimation of Item Parameters

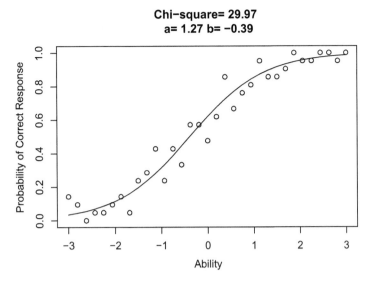

Fig. 3.2 Item characteristic curve fitted to the observed proportions of correct response

fitted item characteristic curve for an item is measured by the chi-square goodness-of-fit index. This index is defined as follows:

$$\chi^2 = \sum_{g=1}^{G} f_g \frac{[p(\theta_g) - P(\theta_g)]^2}{P(\theta_g)Q(\theta_g)}, \tag{3.1}$$

where

G is the number of ability groups,
θ_g is the ability level of group g,
f_g is the number of examinees having ability θ_g
$p(\theta_g)$ is the observed proportion of correct response for group g,
$P(\theta_g)$ is the probability of correct response for group g computed from the item characteristic curve model using the item parameter estimates, and
$Q(\theta_g) = 1 - P(\theta_g)$.

If the value of the obtained index is greater than a criterion value, the item characteristic curve specified by the values of the item parameter estimates does not fit the data. This can be caused by two things. First, an inappropriate item characteristic curve model may have been employed. Second, the values of the observed proportions of correct response are so widely scattered that a good fit, regardless of model, cannot be obtained. In most tests, a few items will yield large values of the chi-square index due to the second reason. However, if many items fail to yield well-fitting item characteristic curves, there may be reason to suspect that an inappropriate model has been employed. In such cases, reanalyzing the test under an alternative model, say the three-parameter model rather than the Rasch model, may yield better results.

In the case of the item shown in Fig. 3.2, the obtained value of the chi-square index was 29.97 and the criterion value was 44.99. Thus the two-parameter model with $b = -0.39$ and $a = 1.27$ was a good fit to the observed proportions of correct response.

The actual maximum likelihood estimation procedure is rather complex mathematically and entails very laborious computations that must be performed for every item in a test.[1] In fact, until computers became widely available, item response theory was not practical because of its heavy computational demands. For present purposes, it is not necessary to go into the details of this procedure. It is sufficient to know that the curve-fitting procedure exists, that it involves a lot of computing, and that the goodness-of-fit of the obtained item characteristic curve can be measured. Because test analysis is done by computer, the computational demands of the item parameter estimation process do not present a major problem today.

3.3 The Group Invariance of Item Parameters

One of the interesting features of item response theory is that the item parameters are not dependent upon the ability level of the examinees responding to the item. Thus, the item parameters are what is known as group invariant. This property of the theory can be described as follows. Assume that you have two groups of examinees drawn from the same population of examinees. The first group has a range of ability scores from -3 to -1 with a mean of -2. The second group has a range of ability scores from $+1$ to $+3$ with a mean of $+2$. Next, the observed proportion of correct response to a given item is computed from the item response data for every ability level within each of the two groups. Then, for the first group, the proportions of correct response are plotted as shown in Fig. 3.3.

The maximum likelihood estimation procedure is then used to fit an item characteristic curve to the data and numerical values of the item parameter estimates, $b(1) = -0.39$ and $a(1) = 1.27$, were obtained. The item characteristic curve defined by these estimates is then plotted over the range of ability encompassed by the first group. This curve is shown in Fig. 3.4.

---

[1] The likelihood, L, that is the probability of observing a set of r_g values from f_g given item parameters is maximized as if it is a function of the item parameters: $L =_{f_g} C_{r_g} \prod_{g=1}^{G} P(\theta_g)^{r_g} Q(\theta_g)^{(f_g - r_g)}$, where C designates combination. Because item parameters that maximize L also maximize the logarithm of L, $\log L$ is used to find the estimates of item parameters. To find the values of item parameter estimates that maximize $\log L$, the Newton-Raphson method can be employed. The partial derivatives as well as the second partial derivatives of $\log L$ with respect to every item parameter for an item are required in the Newton-Raphson method. Using a set of initial values of

3.3 The Group Invariance of Item Parameters

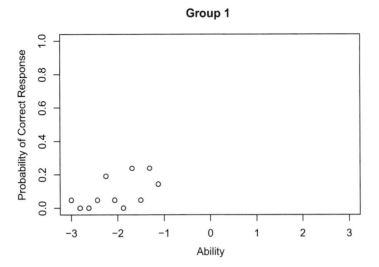

Fig. 3.3 Observed proportions of correct response for group 1

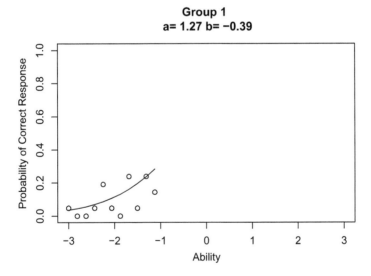

Fig. 3.4 Item characteristic curve fitted to the group 1 data

This process is repeated for the second group. The observed proportions of correct response are shown in Fig. 3.5 and the fitted item characteristic curve with parameter estimates, $b(2) = -0.39$ and $a(2) = 1.27$, is shown in Fig. 3.6.

---

the item parameters the iteration in the Newton-Raphson method will be performed until a stable set of parameter estimates are obtained.

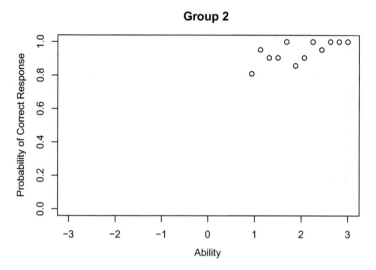

Fig. 3.5 Observed proportions of correct response for group 2

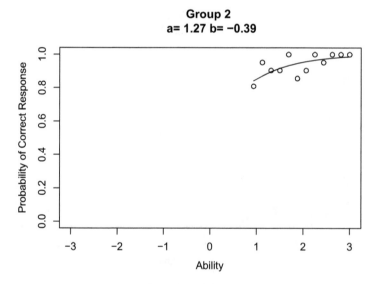

Fig. 3.6 Item characteristic curve fitted to the group 2 data

The result of interest is that under these conditions $b(1) = b(2)$ and $a(1) = a(2)$; that is, the two groups yield the same values of the item parameters. Hence, the item parameters are group invariant. While this result may seem a bit unusual, its validity can be demonstrated easily by considering the process used to fit an item characteristic curve to the observed proportions of correct response. Since the first group had a low average ability (-2), the ability levels spanned by group 1

3.3 The Group Invariance of Item Parameters

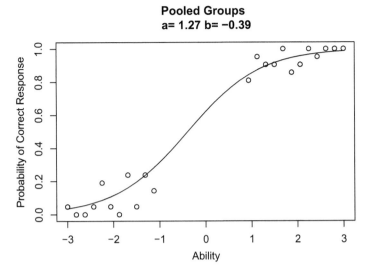

Fig. 3.7 Item characteristic curve fitted to the pooled data

will encompass only a section of the curve, in this case, the lower left tail of the curve. Consequently, the observed proportions of correct response will range from very small to moderate values. When fitting a curve to this data, only the lower tail of the item characteristic curve is involved. For example, see Fig. 3.4. Since group 2 had a high average ability (+2), its observed proportions of correct response range from moderate to very near unity. When fitting an item characteristic curve to this data, only the upper right-hand tail of the curve is involved as was shown in Fig. 3.6. Now, since the same item was administered to both groups, the two curve-fitting processes were dealing with the same underlying item characteristic curve. Consequently, the item parameters yielded by the two analyses should be the same. Figure 3.7 integrates the two previous diagrams into a single representation showing how the same item characteristic curve fits the two sets of proportions of correct response.

The group invariance of the item parameters is a very powerful feature of item response theory. It says that the values of the item parameters are a property of the item, not of the group that responded to the item. Under classical test theory, just the opposite holds. The item difficulty of classical test theory is the overall proportion of correct response to an item for a group of examinees. Thus, if an item with $b = 0$ were responded to by a low-ability group, few of the examinees would get it correct. The classical item difficulty index would yield a low value; say 0.3, as the item difficulty for this group. If the same item were responded to by a high-ability group, most of the examinees would get it correct. The classical item difficulty index would yield a high value; say 0.8, indicating that the item was easy for this group. Clearly, the value of the classical item difficulty index is not group invariant. Because of this, item difficulty, as defined under item response theory is easier to interpret as it has a consistent meaning that is independent of the group used to obtain its value.

Even though the item parameters are group invariant, this does not mean that the numerical values of the item parameter estimates yielded by the maximum likelihood estimation procedure for two groups of examinees, from a common population, taking the same items will always be identical. The obtained numerical values will be subject to variation due to sample size, how well-structured the data is and the goodness-of-fit of the curve to the data. Even though the underlying item parameter values are the same for two samples, the obtained item parameter estimates will vary from sample to sample. Nevertheless, the obtained values should be "in the same ballpark." The result is that in an actual testing situation, the group-invariance principle holds but will not be apparent in the several values of the item parameter estimates obtained for the same items. In addition, the item must be used to measure the same latent trait for both groups. An item's parameters do not retain group invariance when taken out of context; that is, when used to measure a different latent trait, with examinees from a population for which the test is inappropriate, or when the two groups were drawn from two different populations of examinees.

The group invariance of the item parameters also illustrates a basic feature of the item characteristic curve. As stated in earlier chapters, this curve is the relation between the probability of correct response to the item and the ability scale. The invariance principle reflects this since the item parameters are independent of the distribution of examinees over the ability scale. From a practical point of view, this means that the parameters of the total item characteristic curve can be estimated from any segment of the curve. The invariance principle is also one of the bases for test equating under item response theory.

3.4 Computer Session

The purpose of this computer session is twofold. First, it serves to illustrate the fitting of item characteristic curves to the observed proportions of correct response. The computer will generate a set of response data, fit an item characteristic curve to the data under a given model, and then compute the chi-square goodness-of-fit index. This will enable you to see how well the curve-fitting procedure works for a variety of data sets and models. Second, this session shows you that the group invariance of the item parameters holds across models and over a wide range of group definitions. The session allows you to specify the range of ability encompassed by each of two groups of examinees. The computer will generate the observed proportions of correct response for each group and then fit an item characteristic curve to the data. The values of the item parameters are also reported. Thus, you can experiment with various group definitions and observe that the group invariance holds. Example cases and exercises in the next section will be presented for both of these curve-fitting situations.

3.4 Computer Session

3.4.1 Procedures for an Example of Fitting an Item Characteristic Curve to Response Data

The followings are the R command lines to display the observed proportion of correct response for each of 33 ability score levels based on the item characteristic curve model (i.e., mdl) of your choice:

```
> theta <- seq(-3, 3, .1875)
> f <- rep(21, length(theta))
> wb <- round(runif(1,-3,3), 2)
> wa <- round(runif(1,0.2,2.8), 2)
> wc <- round(runif(1,0,.35), 2)
> mdl <- 2
> if (mdl == 1 | mdl == 2) { wc <- 0 }
> if (mdl == 1) { wa <- 1 }
> for (g in 1:length(theta)) {
    P <- wc + (1 - wc) / (1 + exp(-wa * (theta - wb)))
  }
> p <- rbinom(length(theta), f, P) / f
> par(lab=c(7,5,3))
> plot(theta, p, xlim=c(-3,3), ylim=c(0,1),
    xlab="Ability", ylab="Probability of Correct Response")
```

By pressing the enter key in the end of each line, the computer will generate item parameters and the observed proportions of correct response based on the generated item parameters. The screen will be similar in appearance to Fig. 3.1.

From the first line, a sequence of ability score levels are created with -3 as a starting number and 3 as an ending number with an increment of 0.1875. The length of such a sequence is 33. In the second line the R function length was used to obtain the length of the sequence for the ability vector and the number or frequency of examinees for each of the ability score levels is set to be 21. The frequency vector f will contain a set of 21s for the length of 33. Hence, there are 33 groups and each group will have 21 examinees. The total number of examinees is 693.

In the next three lines, item parameters are generated from the uniform distributions. The following lines can also be used:

```
> wb <- runif(1 -3, 3)
> wa <- runin(1, 0.2, 2.8)
> wc <- runif(1, 0, .35)
```

The three arguments in generating a random variate using the uniform distribution (i.e., runif) designate the number of random variates, the minimum of the distribution, and the maximum of the distribution, respectively. For example, a random variate will be generated from a uniform distribution from -3 to 3 and then be defined as an item difficulty parameter wb in the first line. The w in wb indicates the observed proportions will be generated later "with" the item difficulty parameter. Similarly, a random variate will be generated from a uniform distribution from 0.2 to 2.8 and be defined as an item discrimination parameter wa. Also the guessing parameter wc will be generated from a uniform distribution from 0 to 0.35. It will be certainly possible to use some other distributions or some other limits of the uniform distribution to generate the item parameters.

The earlier command lines actually contained the R function round to convert the generated parameters to be rounded to have only two decimal places. The function round can allow us to print out the item parameters in an ordered fashion because they potentially have many decimal places.

The item characteristic curve model is defined with the number of item parameters in the model. The two-parameter model was used in the example by assigning the number of item parameters to be 2. Although the three item parameters were initially generated, based on the model specified, the guessing parameter wc and the item discrimination parameter wa are modified to 0 and 1, respectively. For the Rasch model and the two-parameter model, the guessing parameter will be reassigned to be 0 based on the if conditional statement. For the Rasch model, the item discrimination parameter will be reassigned to be 1 using the if statement.

The vector of the values of the probability of correct response from the specified model given the ability score level θ_g and the item parameters will then be generated and saved as P. The length of P is 33. The random variate based on the binomial distribution with parameters of f and P for each ability level will then be created. The vector that contains the observed proportions of correct response will be obtained and saved as p. The length of p is 33.

In the next line the numbers of ticks in the horizontal and vertical axes are set by the graphical parameter function with its labels argument (i.e., the number of ticks on the horizontal axes is 7 and that on the vertical axes is 5). Using the specified numbers of ticks, the next R function plot will create the plot of the observed proportions of correct response as a function of ability in the graphics window. There are 33 points in the graph.

Assume that you have executed the R command lines listed in the beginning of this subsection. After making the R console window as a current window, the following command lines will now fit an item characteristic curve to the observed proportions of correct response and report the chi-square index and the values of item parameters. The command lines will yield the screen in the graphics window that is similar in appearance to Fig. 3.2:

```
> cs <- 0
> for (g in 1:length(theta)) {
    v <- f[g] * (p[g] - P[g])^2 / (P[g] - P[g]^2)
    cs <- cs + v
  }
> cs <- round(cs, 2)
> if (mdl == 1) {
    maintext <- paste("Chi-square=", cs, "\n", "b=", wb)
  }
> if (mdl == 2) {
    maintext <- paste("Chi-square=",cs,"\n","a=",wa,"b=",wb)
  }
> if (mdl == 3) {
```

3.4 Computer Session

```
          maintext <- paste("Chi-square=", cs, "\n",
             "a=", wa, "b=", wb, "c=", wc)
      }
> par(new="T")
> plot(theta, P, xlim=c(-3,3), ylim=c(0,1), type="l",
      xlab="", ylab="", main=maintext)
```

The first three command lines for cs will obtain the chi-square goodness-of-fit index defined in Eq. (3.1). The first line initializes the starting value of the chi-square index cs to be 0. In the next line, for each ability level g, the term will be calculated and then summed to yield the final value of the chi-square index. The obtained chi-square index will be rounded to yield a value that has two decimal places.

The three if statements prepare the main text to be printed on the plot of the item characteristic curve based on the model specified. Both the value of the chi-square index and the set of item parameters will be printed as the main text on the graph. The plot of the item characteristic curve based on the item parameters will be superimposed onto the existing plot.

Note that the item characteristic curve defined by the estimated values of the item parameters (n.b., in fact, these are not really the estimated values but the parameters that were used to generate the data) is a good fit to the observed proportions of correct response. The obtained value of the chi-square index is usually less than the criterion value from the chi-square distribution. The criterion values at the nominal $\alpha = 0.05$ significance level for the number of the ability score levels of 33 are $\chi^2(32) = 46.19$ for the Rasch model, $\chi^2(31) = 44.99$ for the two-parameter model, and $\chi^2(30) = 43.77$ for the three-parameter model. The exact criterion value of the chi-square distribution for the two-parameter model can be found by using the following quantile function of the chi-square distribution in R:

```
> qchisq(.95, df=31)
```

3.4.2 Procedures for an Example Case Illustrating Group Invariance

Assume that the fitting the item characteristic curve has been performed as in the previous subsection. The followings are the R command lines to display the observed proportions of correct response over the ability score levels for group 1. The lower bound of ability for group 1 is specified with t1l that implies theta-1-lower; the upper bound of ability for group 1 is specified with t1u that implies theta-1-upper in the following command lines:

```
> t1l <- -3
> t1u <- -1
> lowerg1 <- 0
> for (g in 1:length(theta)) {
      if (theta[g] <= t1l) { lowerg1 <- lowerg1 + 1 }
```

```
        }
> upperg1 <- 0
> for (g in 1:length(theta)) {
        if (theta[g] <= t1u) { upperg1 <- upperg1 + 1 }
    }
> theta1 <- theta[lowerg1:upperg1]
> p1 <- p[lowerg1:upperg1]
> if (mdl == 1) { maintext <- paste("Group 1", "\n") }
> if (mdl == 2) { maintext <- paste("Group 1", "\n") }
> if (mdl == 3) { maintext <- paste("Group 1", "\n") }
> plot(theta1, p1, xlim=c(-3,3), ylim=c(0,1),
    xlab="Ability", ylab="Probability of Correct Response",
    main=maintext)
```

By pressing the enter key in the end of each line, the computer will generate a plot that shows the observed proportions of correct response for group 1 in the graphics window. The screen will be similar in appearance to Fig. 3.3.

The lower and upper bounds are specified in the two beginning command lines. The next four lines try to find the appropriate ability score levels contained in the two given bounds of group 1. Then only the ability score levels within the lower and upper bounds of group 1 (i.e., t1) will be selected to yield the plot of the corresponding observed proportions of correct response p1 on the theta scale.

To obtain the plot of an item characteristic curve fitted to the data in the selected ability score levels of group 1, that also reports the values of the item parameters, the following R command lines are executed:

```
> P1 <- P[lowerg1:upperg1]
> if (mdl == 1) {
    maintext <- paste("\n", "b=", wb)
  }
> if (mdl == 2) {
    maintext <- paste("\n", "a=", wa, "b=", wb)
  }
> if (mdl == 3) {
    maintext <- paste("\n", "a=", wa, "b=", wb, "c=", wc)
  }
> par(new="T")
> plot(theta1, P1, xlim=c(-3,3), ylim=c(0,1), type="l",
    xlab="", ylab="", main=maintext)
```

The screen will be similar in appearance to Fig. 3.4.

In order to obtain a graph for group 2 similar in appearance to Fig. 3.5, the required R command lines are as follows:

```
> t2l <- 1
> t2u <- 3
> lowerg2 <- 0
> for (g in 1:length(theta)) {
        if (theta[g] <= t2l) { lowerg2 <- lowerg2 + 1 }
```

3.4 Computer Session

```
     }
> upperg2 <- 0
> for (g in 1:length(theta)) {
     if (theta[g] <= t2u) { upperg2 <- upperg2 + 1 }
  }
> theta2 <- theta[lowerg2:upperg2]
> p2 <- p[lowerg2:upperg2]
> if (mdl == 1) { maintext <- paste("Group 2", "\n") }
> if (mdl == 2) { maintext <- paste("Group 2", "\n") }
> if (mdl == 3) { maintext <- paste("Group 2", "\n") }
> plot(theta2, p2, xlim=c(-3,3), ylim=c(0,1),
     xlab="Ability", ylab="Probability of Correct Response",
     main=maintext)
```

In order to obtain a graph for group 2 similar in appearance to Fig. 3.6, the following R command lines are used:

```
> P2 <- P[lowerg2:upperg2]
> if (mdl == 1) {
     maintext <- paste("\n", "b=", wb)
  }
> if (mdl == 2) {
     maintext <- paste("\n", "a=", wa, "b=", wb)
  }
> if (mdl == 3) {
     maintext <- paste("\n", "a=", wa, "b=", wb, "c=", wc)
  }
> par(new="T")
> plot(theta2, P2, xlim=c(-3,3), ylim=c(0,1), type="l",
     xlab="", ylab="", main=maintext)
```

After defining the two groups and obtaining the plots with respective proportions of correct response together with the fitted item characteristic curves, the following R command lines can be used to obtain a graph for the pooled groups similar in appearance to Fig. 3.7 ultimately:

```
> theta12 <- c(theta1, theta2)
> p12 <- c(p1, p2)
> if (mdl == 1) { maintext <- paste("Pooled Groups","\n") }
> if (mdl == 2) { maintext <- paste("Pooled Groups","\n") }
> if (mdl == 3) { maintext <- paste("Pooled Groups","\n") }
> plot(theta12, p12, xlim=c(-3,3), ylim=c(0,1),
     xlab="Ability", ylab="Probability of Correct Response",
     main=maintext)
> if (mdl == 1) {
     maintext <- paste("\n", "b=", wb)
  }
> if (mdl == 2) {
     maintext <- paste("\n", "a=", wa, "b=", wb)
  }
```

```
> if (mdl == 3) {
    maintext <- paste("\n", "a=", wa, "b=", wb, "c=", wc)
  }
> par(new="T")
> plot(theta, P, xlim=c(-3,3), ylim=c(0,1), type="l",
    xlab="", ylab="", main=maintext)
```

It may be possible to report the value of the chi-square index but not obtained for the pooled data. Only item parameters are reported in the final plot that contains the item characteristic curve for the entire ability scale including the portions of ability score levels outside of the boundary of the ability levels of the two groups. From the graph in the graphics window, the numerical values of the item parameters will be identical to those reported for each of the two groups. From this screen it is clear that the same item characteristic curve has been fitted to both sets of data. This holds even though there was a range of ability scores ($-l$ to $+1$) where there were no observed proportions of correct response to the item.

3.4.3 An R Function for Item Characteristic Curve Fitting

It is possible to create an R function for both generating observed proportions of correct response and fitting an item characteristic curve based on the model selected. Consider the following function named `iccfit`:

```
> iccfit <- function(mdl) {
    theta <- seq(-3, 3, .1875)
    f <- rep(21, length(theta))
    wb <- round(runif(1,-3,3), 2)
    wa <- round(runif(1,0.2,2.8), 2)
    wc <- round(runif(1,0,.35), 2)
    if (mdl == 1 | mdl == 2) { wc <- 0 }
    if (mdl == 1) { wa <- 1 }
    for (g in 1:length(theta)) {
      P <- wc + (1 - wc) / (1 + exp(-wa * (theta - wb)))
    }
    p <- rbinom(length(theta), f, P) / f
    par(lab=c(7,5,3))
    plot(theta, p, xlim=c(-3,3), ylim=c(0,1),
      xlab="Ability", ylab="Probability of Correct Response")
    cs <- 0
    for (g in 1:length(theta)) {
      v <- f[g] * (p[g] - P[g])^2 / (P[g] - P[g]^2)
      cs <- cs + v
    }
    cs <- round(cs, 2)
    if (mdl == 1) {
      maintext <- paste("Chi-square=", cs, "\n", "b=", wb)
    }
```

3.4 Computer Session   49

```
      if (mdl == 2) {
        maintext <- paste("Chi-square=",cs,"\n","a=",wa,"b=",wb)
      }
      if (mdl == 3) {
        maintext <- paste("Chi-square=", cs, "\n",
          "a=", wa, "b=", wb, "c=", wc)
      }
      par(new="T")
      plot(theta, P, xlim=c(-3,3), ylim=c(0,1), type="l",
        xlab="", ylab="", main=maintext)
    }
```

After typing in the function `iccfit` in the R console window, the Rasch model data for the observed proportions of correct response and the corresponding item characteristic curve can be obtained by typing in:

```
> iccfit(1)
```

For the two-parameter model, the command line is:

```
> iccfit(2)
```

For the three-parameter model, the command line is:

```
> iccfit(3)
```

3.4.4 An R Function for the Group Invariance of Item Parameters

It is possible to create an R function for illustrating the group invariance of item parameters. Consider the following function named `groupinv`:

```
> groupinv <- function(mdl, t1l, t1u, t2l, t2u) {
      if (missing(t1l)) t1l <- -3
      if (missing(t1u)) t1u <- -1
      if (missing(t2l)) t2l <- 1
      if (missing(t2u)) t2u <- 3
      theta <- seq(-3, 3, .1875)
      f <- rep(21, length(theta))
      wb <- round(runif(1,-3,3), 2)
      wa <- round(runif(1,0.2,2.8), 2)
      wc <- round(runif(1,0,.35), 2)
      if (mdl == 1 | mdl == 2) { wc <- 0 }
      if (mdl == 1) { wa <- 1 }
      for (g in 1:length(theta)) {
        P <- wc + (1 - wc) / (1 + exp(-wa * (theta - wb)))
      }
      p <- rbinom(length(theta), f, P) / f
      lowerg1 <- 0
```

```
    for (g in 1:length(theta)) {
      if (theta[g] <= t1l) { lowerg1 <- lowerg1 + 1 }
    }
    upperg1 <- 0
    for (g in 1:length(theta)) {
      if (theta[g] <= t1u) { upperg1 <- upperg1 + 1 }
    }
    theta1 <- theta[lowerg1:upperg1]
    p1 <- p[lowerg1:upperg1]
    lowerg2 <- 0
    for (g in 1:length(theta)) {
      if (theta[g] <= t2l) { lowerg2 <- lowerg2 + 1 }
    }
    upperg2 <- 0
    for (g in 1:length(theta)) {
      if (theta[g] <= t2u) { upperg2 <- upperg2 + 1 }
    }
    theta2 <- theta[lowerg2:upperg2]
    p2 <- p[lowerg2:upperg2]
    theta12 <- c(theta1, theta2)
    p12 <- c(p1, p2)
    par(lab=c(7,5,3))
    plot(theta12, p12, xlim=c(-3,3), ylim=c(0,1),
      xlab="Ability", ylab="Probability of Correct Response")
    if (mdl == 1) {
      maintext <- paste("Pooled Groups", "\n", "b=", wb)
    }
    if (mdl == 2) {
      maintext <- paste("Pooled Groups","\n","a=",wa,"b=",wb)
    }
    if (mdl == 3) {
      maintext <- paste("Pooled Groups", "\n",
        "a=", wa, "b=", wb, "c=", wc)
    }
    par(new="T")
    plot(theta, P, xlim=c(-3,3), ylim=c(0,1), type="l",
      xlab="", ylab="", main=maintext)
}
```

After typing in the function groupinv in the R console window, the Rasch model data for illustrating the group invariance of the item difficulty parameter can be done by typing in:

> groupinv(1)

Note that the default values of the lower and upper bounds of group 1 are -3 and -1, respectively, and that those of group 2 are 1 and 3. The above command line is equivalent to:

> groupinv(1, -3, -1, 1, 3)

3.5 Exercises

It is certainly possible to use different sets of the lower and upper bounds for group 1 and group 2 as long as the value of the upper bound is strictly greater than that of the lower bound with a reasonable width on the ability scale. If the default boundary values are used for the two-parameter model, the illustration can be done with:

```
> groupinv(2)
```

For the three-parameter model with the default boundary values, the command line is:

```
> groupinv(3)
```

3.5 Exercises

For the following three exercises, it is assumed that you have defined the function iccfit by typing it in the R console window. These exercises enable you to develop a sense of how well the obtained item characteristic curves fit the observed proportions of correct response. The criterion value of the chi-square index will be based on the item characteristic curve model selected (i.e., 46.19 for the Rasch model, 44.99 for the two-parameter model, and 43.77 for the three-parameter model). This criterion value actually depends upon the number of ability score levels used and the number of parameters estimated. Thus, it will vary from situation to situation. For present purposes it will be sufficient to use these criterion values based on the 33 ability score levels.

1. Repeat several times the fitting of an item characteristic curve to the response data generated using the Rasch model.
2. Repeat several times the fitting of an item characteristic curve to the response data generated using the two-parameter model.
3. Repeat several times the fitting of an item characteristic curve to the response data generated using the three-parameter model.

3.5.1 Further Exercises

The following exercises enable you to examine the group-invariance principle under all three item characteristic curve models and for a variety of group definitions. You may use the function groupinv by executing it in the R console window.

1. Under the two-parameter model (i.e., mdl <- 2), set the following ability bounds:

    Group 1—The lower bound is -2, and the upper bound is $+1$.
    Group 2—The lower bound is -1, and the upper bound is $+2$.

Generate the five display graphs (e.g., Figs. 3.3, 3.4, 3.5, 3.6 and 3.7) sequentially in the graphics window for this example. In case the function `groupinv` is employed that will yield only a graph that is similar in appearance to Fig. 3.7, the following command line can be used:

```
> groupinv(2, -2, 1, -1, 2)
```

2. Under the Rasch model (i.e., `mdl <- 1`), set the following ability bounds:

   Group 1—The lower bound is −3, and the upper bound is −1.
   Group 2—The lower bound is +1, and the upper bound is +3.

   Study the resulting display screens. Then try:

   Group 1—The lower bound is −2, and the upper bound is +1.
   Group 2—The lower bound is −1, and the upper bound is +2.

3. Under the three-parameter model (i.e., `mdl <- 3`), set the following ability bounds:

   Group 1—The lower bound is −3, and the upper bound is −1.
   Group 2—The lower bound is +1, and the upper bound is +3.

   Study the resulting display screens. Then try:

   Group 1—The lower bound is −2, and the upper bound is +1.
   Group 2—The lower bound is −1, and the upper bound is +2.

4. Now experiment with various combinations of overlapping and nonoverlapping ability groups in conjunction with each of the three item characteristic curve models.

3.6 Things to Notice

1. Under all three models, the item characteristic curve based upon the estimated item parameters was usually a good overall fit to the observed proportions of correct response. In these exercises, this is more of a function of the manner in which the observed proportions of correct response were generated than of some intrinsic property of the item characteristic curve models. However, in most well-constructed tests the majority of item characteristic curves specified by the item parameter estimates will fit the data. The lack of fit usually indicates that that item needs to be studied and perhaps rewritten or discarded.
2. When two groups are employed, the same item characteristic curve will be fitted regardless of the range of ability encompassed by each of the two groups.
3. The distribution of examinees over the range of abilities for a group was not considered. Only the ability levels are of interest. How many examinees have each of these levels does not affect the group-invariance property.

3.6 Things to Notice

4. If two groups of examinees are separated along the ability scale and the item has positive discrimination, the low-ability group involves the lower left tail of the item characteristic curve. The high-ability group involves the upper right tail.
5. The item parameters were group invariant whether or not the ability ranges of the two groups overlapped. Thus, overlap is not a consideration.
6. If you were brave enough to define group 1 as the high-ability group and group 2 as the low-ability group, you would have discovered that it made no difference as to which group was the high-ability group. Thus, group labeling is not a consideration.
7. The group-invariance principle holds for all three item characteristic curve models.
8. It is important to recognize that whenever item response data is used, the obtained item parameter estimates are subject to sampling variation. As a result, the same test administered to several groups of students will not yield the same numerical values for the item parameter estimates each time. However, this does not imply that the group-invariance principle is invalid. It simply means that the principle is more difficult to observe in real data.

Chapter 4
The Test Characteristic Curve

4.1 Introduction

Item response theory is based upon the individual items of a test, and up to this point the chapters have dealt with the items one at a time. Now, attention will be given to dealing with all the items in a test at once. When scoring a test, the response made by an examinee to each item is dichotomously scored. A correct response is given a score of 1 and an incorrect response a score of 0; the examinee's raw test score is obtained by adding up the item scores. This raw test score will always be an integer number and will range from 0 to J, where J is the number of items in the test. If examinees were to take the test again, assuming they did not remember how they previously answered the items, a different raw test score would be obtained. Hypothetically, an examinee could take the test a great many times and obtain a variety of test scores. One would anticipate that these scores would cluster themselves around some average value. In measurement theory, this value is known as the true score and its definition depends upon the particular measurement theory. In item response theory, the definition of a true score according to D.N. Lawley is used.

4.2 A True Score

The formula for a true score is given in Eq. (4.1) below:

$$\text{TS}_i = \sum_{j=1}^{J} P_j(\theta_i), \tag{4.1}$$

where

TS$_i$ is the true score for examinees with ability level θ_i,
j denotes an item, $j = 1, \ldots, J$, and
$P_j(\theta_i)$ depends on the particular item characteristic curve model employed.

The task at hand is to calculate the true score for those examinees having a given ability level. To illustrate this, the probability of correct response for each item in a four-item test will be calculated at an ability level of 1.0. This can be done using the formula for the two-parameter model and the procedures given in Chap. 2. The two-parameter model for item j is now defined as:

$$P_j(\theta) = \frac{1}{1 + \exp[-a_j(\theta - b_j)]} \equiv \frac{1}{1 + e^{-a_j(\theta - b_j)}}.$$

The item discrimination parameters and the item difficulty parameters for the four-item test are as follows:

Item 1: $a_1 = 0.5$ and $b_1 = -1.0$
Item 2: $a_2 = 1.2$ and $b_2 = 0.75$
Item 3: $a_3 = 0.8$ and $b_3 = 0.0$
Item 4: $a_4 = 1.0$ and $b_4 = 0.5$

With $\theta_i = 1.0$, the probability of correct response for each item under the two-parameter model can be obtained as:

$$P_1(1.0) = \frac{1}{1 + \exp[-0.5(1.0 - (-1.0))]} = 0.731058578$$
$$P_2(1.0) = \frac{1}{1 + \exp[-1.2(1.0 - 0.75)]} = 0.574442516$$
$$P_3(1.0) = \frac{1}{1 + \exp[-0.8(1.0 - 0.0)]} = 0.689974481$$
$$P_4(1.0) = \frac{1}{1 + \exp[-1.0(1.0 - 0.5)]} = 0.622459331$$

The item characteristic curve for item 1 is presented in Fig. 4.1. To make the process a bit clearer, the dashed lines on the figure show the relation between the value of $\theta_i = 1.0$ and $P_1(1.0)$ on the item characteristic curve. Figures 4.2, 4.3 and 4.4 present the respective item characteristic curves for items 2–4.

Now, to get the true score at $\theta_i = 1.0$, the probabilities of correct response are summed over the four items:

TS$_i = 0.731058578 + 0.574442516 + 0.689974481 + 0.622459331 = 2.617934906$

Thus, for examinees having an underlying ability of 1.0, their true score on this test would be 2.62. This score is intuitively reasonable because, at an ability score of 1.0, each of the item characteristic curves is above 0.5 and the sum of the probabilities would be large. While no individual examinee would actually get this score, it is the theoretical average of all the raw test scores that examinees of ability 1.0 would get

4.2 A True Score

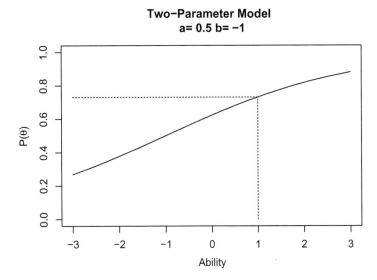

Fig. 4.1 Item characteristic curve for item 1

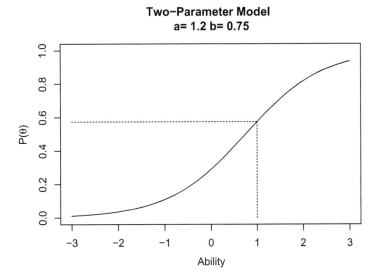

Fig. 4.2 Item characteristic curve for item 2

on this test of four items had they taken the test a large number of times. Actual tests would contain many more items than four, but the true score would be obtained in the same manner.

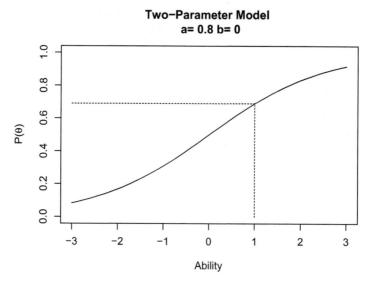

Fig. 4.3 Item characteristic curve for item 3

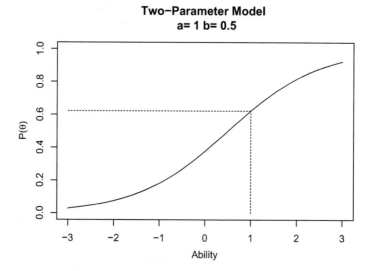

Fig. 4.4 Item characteristic curve for item 4

4.3 The Test Characteristic Curve

The calculations performed above were for a single point on the ability scale. This true score computation can be performed for any point along the ability scale from negative infinity to positive infinity. The corresponding true scores then could be

4.3 The Test Characteristic Curve

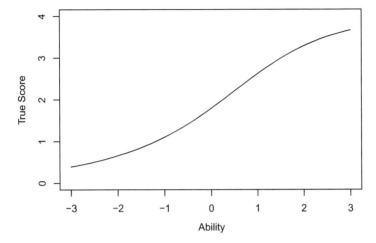

Fig. 4.5 Test characteristic curve

plotted as a function of ability. The vertical axis would be the true scores and would range from zero to the number of items in the test. The horizontal axis would be the ability scale. These plotted scores will form a smooth curve and this curve is the test characteristic curve. Figure 4.5 depicts a typical test characteristic curve for a test containing four items.

The test characteristic curve is the functional relation between the true score and the ability scale. Given any ability level, the corresponding true score can be found via the test characteristic curve. For example, in Fig. 4.5 draw a vertical line at an ability score of 1.0 upward until the test characteristic curve is intersected. Then, draw a horizontal line to the left until it intersects the true score scale. This line yields a true score of 2.62 for an ability score of 1.0.

When the Rasch model or the two-parameter model is used for the J items in a test, the left tail of the test characteristic curve approaches zero as the ability score approaches negative infinity; its upper tail approaches the number of items in the test as the ability score approaches positive infinity. The implication of this is that under these two models a true score of zero corresponds to an ability score of negative infinity and a true score of J corresponds to an ability level of positive infinity. When the three-parameter model is used for the J items in a test, the lower tail of the test characteristic curve approaches the sum of the guessing parameters for the test items rather than zero. This reflects the fact that under this model very low-ability examinees can get a test score greater than zero simply by guessing. The upper tail of the test characteristic curve still approaches the number of items in the test. Hence, a true score of J corresponds to an ability of positive infinity under all three item characteristic curve models.

The primary role of the test characteristic curve in item response theory is to provide a means of transforming ability scores to true scores. This becomes of interest in practical situations where the user of the test may not be able to interpret an ability score. By transforming the ability score into a true score, the user is given a number that relates to the number of items in the test. This number is in a more familiar frame of reference and the user is able to interpret it. However, those familiar with item response theory, such as you, can interpret the ability score directly. The test characteristic curve also plays an important role in the procedures for equating tests.

The general form of the test characteristic curve is that of a monotonically increasing function. In some cases it has a rather smooth S-shape similar to an item characteristic curve. In other cases it will increase smoothly, then have a small plateau before increasing again. However, in all cases it will be asymptotic to a value of J in the upper tail. The shape of the test characteristic curve depends upon a number of factors including the number of items, the item characteristic curve model employed, and the values of the item parameters. Because of this, there is no explicit formula, other than Eq. (4.1), for the test characteristic curve as there was for the item characteristic curve. The only way one can obtain the test characteristic curve is to evaluate the probability of correct response at each ability level for all the items in the test using a given item characteristic curve model. Once these probabilities are obtained, they are summed at each ability level and then the sums are plotted to get the test characteristic curve.

It is very important to understand that the form of the test characteristic curve does not depend upon the frequency distribution of the examinees' ability scores over the ability scale. In this respect, the test characteristic curve is similar to the item characteristic curve. Both are functional relations between two scales and do not depend upon the distribution of scores over the scales.

The test characteristic curve can be interpreted in roughly the same terms as was the item characteristic curve. The ability level corresponding to the mid-true score (i.e., $J/2$) locates the test along the ability scale. The general slope of the test characteristic curve is related to how the value of the true score depends upon the ability level. In some situations, the test characteristic curve is nearly a straight line over much of the ability scale. In most tests, however, the test characteristic curve is nonlinear and the slope is only descriptive for a reduced range of ability levels. Since there is no explicit formula for the test characteristic curve, there are no simple parameters for the curve. The mid true score defines the test difficulty in numerical terms, but the slope of the test characteristic curve is best defined in verbal terms. For most interpretive uses, these two descriptors are sufficient for discussing a test characteristic curve that has been plotted and can be visually inspected.

4.4 Computer Session

This session has several purposes. The first is to show the form of the test characteristic curve and have you develop a feel for how true scores and ability are related in various tests. The second is to show the dependence of the form of the test characteristic curve upon the mix of item parameters occurring in the J items of

4.4 Computer Session

the test. The computer session allows you to set the values of the item parameters for the J items of the test and the computer will plot the resultant test characteristic curve.

4.4.1 Procedures for an Example Case

This example will illustrate how to obtain a test characteristic curve for a small test. The followings are the R command lines to obtain true scores for a four-item test along the ability scale and display the test characteristic curve based on the two-parameter model on the screen:

```
> b <- c(-1.0, 0.75, 0.0, 0.5)
> a <- c(0.5, 1.2, 0.8, 0.75)
> theta <- seq(-3, 3, .1)
> ts <- rep(0, length(theta))
> J <- length(b)
> for (j in 1:J) {
    P <-  1 / (1 + exp(-a[j] * (theta - b[j])))
    ts <- ts + P
  }
> plot(theta, ts, type="l", xlim=c(-3,3), ylim=c(0,J),
    xlab="Ability", ylab="True Score")
```

By pressing the enter key in the end of each line, the computer will use the set of item parameters to obtain the true scores a long the ability scale. The screen in the graphics window will be similar in appearance to Fig. 4.5.

With the first line, a vector that contains item difficulty parameters for four items is created. With the second line, a vector that contains item discrimination parameters for four item is created. Then, the sequence of ability levels is created with -3 as a starting number and 3 as an ending number with an increment of 0.1. The length of such a sequence is 61. The true score vector $\texttt{ts}$ is initialized in the fourth line to have the values of 0 for the same length as ability. The total number of items was defined as J. In the $\texttt{for}$ loop, the probability of correct response is to be calculated with each set of item parameters over the ability levels and to be added to the provisional true score vector. After the looping there are 33 values of true scores given the ability levels. These values are plotted as the test characteristic curve in the last command line.

A test characteristic curve for a test with a different number of items under the two-parameter model can be easily constructed using similar R command lines but specifying different values of the vectors of item parameters. For example, by replacing the first two command lines, we can obtain Fig. 4.6 for a five-item test:

```
> b <- c(-2.0, -1.0, 0.0, 1.0, 2.0)
> a <- c(0.5, 0.75, 1.0, 0.75, 0.5)
> theta <- seq(-3, 3, .1)
```

Fig. 4.6 Test characteristic curve for the five-item test

```
> ts <- rep(0, length(theta))
> J <- length(b)
> for (j in 1:J) {
    P <-  1 / (1 + exp(-a[j] * (theta - b[j])))
    ts <- ts + P
  }
> plot(theta, ts, type="l", xlim=c(-3,3), ylim=c(0,J),
    xlab="Ability", ylab="True Score",
    main="Test Characteristic Curve")
```

In the display, the range of true scores has been bounded by zero and the number of items in the test. Note again that the maximum value of a true score is the number of items in the test, that is, J. Although the command lines generate similarly looking graphs to show test characteristic curves based upon different numbers of items, the true score range on the vertical axis will be accordingly changed.

It should be noted that because the ability range has been restricted arbitrarily to -3 to $+3$, the test characteristic curve may not get very close to either its theoretical upper and lower limits in the plotted curves. You should keep in mind that, had the ability scale gone from negative infinity to positive infinity, the theoretical limits of the true scores would have been seen.

The test characteristic curve for this five-item test is very close to a straight line over the ability range from -2 to $+2$. Outside these values it curves slightly. Thus, there is almost a linear relationship here between ability and true scores having a slope of about 0.5. The small slope reflects the low to moderate values of the item discrimination parameters. The mid true score of 2.5 occurs at an ability level of zero, which reflects the average value of the bs.

4.4.2 An R Function for Test Characteristic Curves

It is possible to create an R function for plotting a test characteristic curve given a set of item parameters. Consider the following function named `tcc`:

```
> tcc <- function(b, a, c) {
    J <- length(b)
    if (missing(c)) c <- rep(0, J)
    if (missing(a)) a <- rep(1, J)
    theta <- seq(-3, 3, .1)
    ts <- rep(0, length(theta))
    for (j in 1:J) {
      P <-  c[j] + (1 - c[j]) / (1 + exp(-a[j]*(theta-b[j])))
      ts <- ts + P
    }
    plot(theta, ts, type="l", xlim=c(-3,3), ylim=c(0,J),
      xlab="Ability", ylab="True Score",
      main="Test Characteristic Curve")
  }
```

Note that use of the function `tcc` requires the vectors of item parameters. After typing in the function in the R console window, the vectors of item parameters should be constructed before the function is executed.

For the Rasch model, a test characteristic curve for a five-item test can be obtained with the function `tcc` by defining a set of item difficulty parameters as:

```
> b <- c(-2.0, -1.0, 0.0, 1.0, 2.0)
> tcc(b)
```

For the two-parameter model, a test characteristic curve for a five-item test can be obtained after defining the respective vectors of item difficulty parameters and item discrimination parameters; for example:

```
> b <- c(-2.0, -1.0, 0.0, 1.0, 2.0)
> a <- c(0.5, 0.75, 1.0, 0.75, 0.5)
> tcc(b, a)
```

For the three-parameter model, a test characteristic curve for a five-item test can be obtained after defining the three vectors of item parameters; for example:

```
> b <- c(-2.0, -1.0, 0.0, 1.0, 2.0)
> a <- c(0.5, 0.75, 1.0, 0.75, 0.5)
> c <- c(.2, .2, .2, .2, .2)
> tcc(b, a, c)
```

It can be noted that the order of item parameter vectors in the function `tcc` as a set of arguments is b, a, and c. The item parameter vectors, however, can be defined with a different order in the R command line by exactly specifying the arguments of the function. The last command line in the three-parameter model example is equivalent to:

```
> tcc(a=a, b=b, c=c)
```

4.5 Exercises

For the following exercises, it is assumed that you have defined the function `tcc` by typing it in the R console window.

4.5.1 Using the Rasch Model

1. Obtain the test characteristic curve for a test with ten items by setting all item difficulty parameters to a value of $b = 0.0$. The test characteristic curve in this case looks just like an item characteristic curve for an item with $b = 0.0$. The vertical axis ranges from 0 to 10. The mid true score occurs at an ability level of zero.
2. Obtain the test characteristic curve for a test with ten items using the following item difficulty parameters:

$$b_1 = -3.0 \quad b_2 = -2.5 \quad b_3 = -2.0 \quad b_4 = -1.5 \quad b_5 = -1.0$$
$$b_6 = -0.5 \quad b_7 = 0.0 \quad b_8 = 0.5 \quad b_9 = 1.0 \quad b_{10} = 1.5$$

The resulting test characteristic curve has a nearly linear section from an ability level of -3 to $+1$. After this point it bends over slightly as it approaches a score of J. The mid true score of 5 corresponds to an ability level of -0.6.
3. Obtain the test characteristic curve for a test with ten items using the following item difficulty parameters:

$$b_1 = -0.8 \quad b_2 = -0.5 \quad b_3 = -0.5 \quad b_4 = 0.0 \quad b_5 = 0.0$$
$$b_6 = 0.0 \quad b_7 = 0.5 \quad b_8 = 0.5 \quad b_9 = 0.5 \quad b_{10} = 0.8$$

The test characteristic curve has a well-defined S-shape much like an item characteristic curve. Only the section near an ability level of zero is linear. The mid true score of 5 corresponds to an ability score of 0.5.

4.5.2 Using the Two-Parameter Model

1. Obtain the test characteristic curve for a test with five items using the following item difficulty and item discrimination parameters:

$$b_1 = 0.0 \quad b_2 = 0.0 \quad b_3 = 0.0 \quad b_4 = 0.0 \quad b_5 = 0.0$$
$$a_1 = 0.4 \quad a_2 = 0.8 \quad a_3 = 0.4 \quad a_4 = 0.8 \quad a_5 = 0.4$$

4.5 Exercises

The test characteristic curve is nearly a straight line with a rather shallow slope reflecting the low to moderate values of a. The mid true score of 2.5 occurs, as expected, at an ability level of 0.0.

2. Obtain the test characteristic curve for a test with five items using the following item difficulty and item discrimination parameters:

$$b_1 = 1.0 \quad b_2 = 1.0 \quad b_3 = 1.0 \quad b_4 = 1.0 \quad b_5 = 1.0$$
$$a_1 = 1.6 \quad a_2 = 1.9 \quad a_3 = 1.6 \quad a_4 = 1.9 \quad a_5 = 1.6$$

The majority of the test characteristic curve is compressed into a rather small section of the ability scale. Up to an ability level of -1 the true score is nearly zero. Beyond an ability level of 2.5, the true score is approaching a value of J. Between these two limits, the curve has a definite S-shape and the steep slope reflects the high values of the item discrimination parameters. The mid true score of 2.5 occurs at an ability level of 1.0. Notice how the difference in average level of item discrimination parameters in these last two problems shows in the difference of the steepness of the two curves.

3. Obtain the test characteristic curve for a test with five items using the following item difficulty and item discrimination parameters:

$$b_1 = -2.0 \quad b_2 = -1.5 \quad b_3 = -1.0 \quad b_4 = -0.5 \quad b_5 = 0.0$$
$$a_1 = 0.4 \quad a_2 = 1.7 \quad a_3 = 0.9 \quad a_4 = 1.6 \quad a_5 = 0.8$$

The test characteristic curve has a moderate S-shape and has a mid true score of 2.5 at an ability level of -0.8, which is not the average value of b but is close to it.

4.5.3 Using the Three-Parameter Model

1. Obtain the test characteristic curve for a test with five items using the following item difficulty, item discrimination, and guessing parameters:

$$b_1 = 1.0 \quad b_2 = 1.2 \quad b_3 = 1.5 \quad b_4 = 1.8 \quad b_5 = 2.0$$
$$a_1 = 1.2 \quad a_2 = 0.9 \quad a_3 = 1.0 \quad a_4 = 1.5 \quad a_5 = 0.6$$
$$c_1 = 0.25 \quad c_2 = 0.20 \quad c_3 = 0.25 \quad c_4 = 0.20 \quad c_5 = 0.30$$

The test characteristic curve has a very long lower tail that levels out just above a true score of 1.2, which is the sum of the values of the parameter c for the five items. Because of the long lower tail, there is very little change in true scores from an ability level of -3.0 to 0.0. Above an ability level of zero, the curve slopes up and begins to approach a true score of J. The mid true score of 2.5

corresponds to an ability level of about 0.5. Thus, the test functions among the high-ability examinees even though, due to guessing, low-ability examinees have true scores just above 1.2.

2. Obtain the test characteristic curve for a test with ten items using the following item difficulty, item discrimination, and guessing parameters:

$b_1 = 2.34 \quad b_2 = -1.09 \quad b_3 = -1.65 \quad b_4 = -0.40 \quad b_5 = 2.90$
$b_6 = -1.54 \quad b_7 = -1.52 \quad b_8 = -1.81 \quad b_9 = -0.63 \quad b_{10} = -2.45$
$a_1 = 1.90 \quad a_2 = 1.64 \quad a_3 = 2.27 \quad a_4 = 0.94 \quad a_5 = 1.83$
$a_6 = 2.67 \quad a_7 = 2.01 \quad a_8 = 1.98 \quad a_9 = 0.92 \quad a_{10} = 2.54$
$c_1 = 0.30 \quad c_2 = 0.30 \quad c_3 = 0.07 \quad c_4 = 0.12 \quad c_5 = 0.16$
$c_6 = 0.27 \quad c_7 = 0.17 \quad c_8 = 0.27 \quad c_9 = 0.28 \quad c_{10} = 0.07$

Compared to the previous test characteristic curves, this one is quite different. The curve goes up sharply from an ability level of -3 to -1. Then it changes quite rapidly into a rather flat line that slowly approaches a value of J. The mid true score of 5.0 is at an ability level of -1.5, indicating that the test functions among the low-ability examinees. This reflects the fact that all but two of the items had a negative value of the parameter b.

3. These are for some exploratory exercises:

   (a) Select the number of items of your choice (e.g., $J = 10$).
   (b) Select the item characteristic curve model of your choice (e.g., the three-parameter model).
   (c) Set the vectors of item parameters with the values of your choice.
   (d) When the test characteristic curve appears on the screen, try to relate the shape of the curve and the set of item parameters you used.
   (e) Using the same number of items and the same item characteristic model, set the item parameters to the values of your choice that are different from the earlier set.
   (f) Try to display the new test characteristic curve on the same graph as the previous curve.
   (g) Repeat the process until you can predict what effect the changed item parameters will have on the form of the test characteristic curve.

4.6 Things to Notice

1. Relation of the true score and the ability level:

   (a) Given an ability level, the corresponding true score can be found via the test characteristic curve.
   (b) Given a true score, the corresponding ability level can be found via the test characteristic curve.
   (c) Both the true scores and ability are continuous variables.

2. Shape of the test characteristic curve:
   (a) When $J = 1$, the true score ranges from 0 to 1 and the shape of the test characteristic curve is identical to that of the item characteristic curve for the single item.
   (b) The test characteristic curve does not always look like an item characteristic curve. It can have regions of varying steepness and plateaus. Such curves reflect a mixture of item parameter values having a large range of values.
   (c) The ability level at which the mid true score ($J/2$) occurs depends primarily upon the average value of the item difficulty parameters and is an indicator of where the test functions on the ability scale.
   (d) When the values of the item difficulty parameters have a limited range, the steepness of the test characteristic curve depends primarily upon the average value of the item discrimination parameters. When the values of the item difficulty parameters are spread widely over the ability scale, the steepness of the test characteristic curve will be reduced even though the values of the item discriminations remain the same.
   (e) Under the three-parameter model the lower limit of the true scores is the sum of the values of the parameter c for the J items of the test.
   (f) The shape of the test characteristic curve depends upon the number of items in the test, the item characteristic curve model and the mix of values of the item parameters possessed by the J items in the test.

3. It would be possible to construct a test characteristic curve that decreases as ability increases. To do so would require items with negative item discrimination for the correct response to the items. Such a test would not be considered to be a good test because the higher an examinee's ability level, the lower the score expected for the examinee.

Chapter 5
Estimating an Examinee's Ability

5.1 Introduction

Under item response theory, the primary purpose for administering a test to an examinee is to locate that person on the ability scale. If such an ability measure can be obtained for each person taking the test, two goals can be achieved. First, the examinee can be evaluated in terms of how much underlying ability he or she possesses. Second, comparisons among examinees can be made for purposes of assigning grades, awarding scholarships, etc. Thus, the focus of this chapter is upon the examinees and the procedures for estimating an ability score (parameter) for an examinee.

The test used to measure an unknown latent trait will consist of J items each of which measures some facet of the trait. In the previous chapters, dealing with item parameters and their estimation, it was assumed that the ability parameter of each examinee was known. Conversely, to estimate an examinee's unknown ability parameter, it will be assumed that the numerical values of the parameters of the test items are known. A direct consequence of this assumption is that the metric of the ability scale will be the same as the metric of the known item parameters. When the test is taken, an examinee will respond to each of the J items in the test and the responses will be dichotomously scored. The result will be a score of either 1 or 0 for each item in the test. It is common practice to refer to the item score of 1 or 0 as the examinee's item response. Thus, the list of 1s and 0s for the J items is called the examinee's item response vector. The task at hand is to use this item response vector and the known item parameters to estimate the examinee's unknown ability parameter.

5.2 Ability Estimation Procedures

Under item response theory, maximum likelihood procedures are used to estimate an examinee's ability. As was the case for item parameter estimation, this procedure is an iterative process. It begins with some a priori value for the ability of the examinee and the known values of the item parameters. These are used to compute the probability of correct response to each item for that examinee. Then an adjustment to the ability estimate is obtained that improves the agreement of the computed probabilities with the examinee's item response vector. The process is repeated until the adjustment becomes small enough that the change in the estimated ability is negligible. The result is an estimate of the examinee's ability parameter. This process is then repeated separately for each examinee taking the test. In Chap. 7, a procedure will be presented through which the ability levels of all examinees are estimated simultaneously. However, this procedure is based upon an approach that treats each examinee separately. Hence, the basic issue is how the ability of a single examinee can be estimated.

The ability estimation equation for the two-parameter model is shown below:

$$\hat{\theta}_{s+1} = \hat{\theta}_s - \frac{\sum_{j=1}^{J} a_j [u_j - P_j(\hat{\theta}_s)]}{-\sum_{j=1}^{J} a_j^2 P_j(\hat{\theta}_s) Q_j(\hat{\theta}_s)}, \tag{5.1}$$

where

$\hat{\theta}_s$ is the provisional, estimated ability of the examinee within iteration s,
a_j is the item discrimination parameter of item j, $j = 1, 2, \ldots, J$,
u_j is the response made by the examinee to item j, where

$u_j = 1$ for a correct response and
$u_j = 0$ for an incorrect response,

$P_j(\hat{\theta}_s)$ is the probability of correct response to item j, under the given item characteristic curve model, at ability level $\hat{\theta}$ within iteration s, and
$Q_j(\hat{\theta}_s) = 1 - P_j(\hat{\theta}_s)$ is the probability of incorrect response to item j, under the given item characteristic curve model, at ability level $\hat{\theta}$ within iteration s.

The equation has a rather simple explanation. Initially, the $\hat{\theta}_s$ on the right side of the equal sign is set to some arbitrary value, such as 1. The probability of correct response to each of the J items in the test is calculated at this ability level using the known item parameters in the given item characteristic curve model. Then the second term to the right of the equal sign is evaluated. This is the adjustment term, denoted by $\Delta \hat{\theta}_s$ as shown in

5.2 Ability Estimation Procedures

$$\hat{\theta}_{s+1} = \hat{\theta}_s - \Delta\hat{\theta}_s.$$

The value of $\hat{\theta}_{s+1}$ on the left side of the equal sign is obtained by subtracting $\Delta\hat{\theta}_s$ from $\hat{\theta}_s$. This value, $\hat{\theta}_{s+1}$, becomes $\hat{\theta}_s$ in the next iteration. The numerator of the adjustment term contains the essence of the procedure. Notice that $[u_j - P_j(\hat{\theta}_s)]$ is the difference between the examinee's item response to item j and the probability of correct response at an ability level of $\hat{\theta}_s$. Now, as the ability estimate gets closer to the examinee's ability, the sum of the differences between u_j and $P_j(\hat{\theta}_s)$ gets smaller. Thus, the goal is to find the ability estimate yielding values of $P_j(\hat{\theta}_s)$ for all items simultaneously that minimizes this sum. When this happens, the $\Delta\hat{\theta}_s$ term becomes as small as possible and the value of $\hat{\theta}_{s+1}$ will not change from iteration to iteration. This final value of $\hat{\theta}_{s+1}$ is then used as the examinee's estimated ability. The ability estimate will be in the same metric as the numerical values of the item difficulty parameters. One nice feature of Eq. (5.1) is that it can be used with all three item characteristic curve models, although the three-parameter model requires a slight modification.

A three-item test will be used to illustrate the ability estimation process. Under the two-parameter model, the known item parameters are:

$$b_1 = -1.0 \quad b_2 = 0.0 \quad b_3 = 1.0$$
$$a_1 = 1.0 \quad a_2 = 1.2 \quad a_3 = 0.8$$

The examinee's item responses were:

$$u_1 = 1 \quad u_2 = 0 \quad u_3 = 1$$

The a priori estimate of the examinee's ability is set to $\hat{\theta}_s = 1.0$. Substituting these values to the two-parameter model,

$$P_j(\hat{\theta}_s) = \frac{1}{1 + \exp[-a_j(\hat{\theta}_s - b_j)]},$$

the first iteration yields:

Item j	u_j	$P_j(\hat{\theta}_s)$	$Q_j(\hat{\theta}_s)$	$a_j[u_j - P_j(\hat{\theta}_s)]$	$a_j^2 P_j(\hat{\theta}_s)Q_j(\hat{\theta}_s)$
1	1	0.88	0.12	0.1192	0.1050
2	0	0.77	0.23	−0.9222	0.2562
3	1	0.50	0.50	0.4000	0.1600
Sum				−0.4030	0.5212

$\Delta\hat{\theta}_s = -0.4030/(-0.5212) = 0.7733$, and $\hat{\theta}_{s+1} = 1.0 - 0.7733 = 0.2267$. This value $\hat{\theta}_{s+1}$ becomes $\hat{\theta}_s$ in the next iteration.

The second iteration yields:

Item j	u_j	$P_j(\hat{\theta}_s)$	$Q_j(\hat{\theta}_s)$	$a_j[u_j - P_j(\hat{\theta}_s)]$	$a_j^2 P_j(\hat{\theta}_s)Q_j(\hat{\theta}_s)$
1	1	0.77	0.23	0.2268	0.1753
2	0	0.57	0.43	−0.6811	0.3534
3	1	0.35	0.65	0.5199	0.1456
Sum				0.0656	0.6744

$\Delta\hat{\theta}_s = 0.0656/(-0.6744) = -0.0972$, and $\hat{\theta}_{s+1} = 0.2267 - (-0.0972) = 0.3239$. Again, this value $\hat{\theta}_{s+1}$ becomes $\hat{\theta}_s$ in the next iteration.

The third iteration yields:

Item j	u_j	$P_j(\hat{\theta}_s)$	$Q_j(\hat{\theta}_s)$	$a_j[u_j - P_j(\hat{\theta}_s)]$	$a_j^2 P_j(\hat{\theta}_s)Q_j(\hat{\theta}_s)$
1	1	0.79	0.21	0.2102	0.1660
2	0	0.60	0.40	−0.7152	0.3467
3	1	0.37	0.63	0.5056	0.1488
Sum				0.0006	0.6616

$\Delta\hat{\theta}_s = 0.0006/(-0.6616) = -0.0009$, and $\hat{\theta}_{s+1} = 0.3239 - (-0.0009) = 0.3248$. At this point, the process is terminated as the absolute value of the adjustment (0.0009) is very small. Thus, the examinee's estimated ability is 0.32.

Unfortunately, there is no way to know the examinee's actual ability parameter. The best one can do is to estimate it. However, this does not prevent us from conceptualizing such a parameter. Fortunately, one can obtain a standard error of the estimated ability that provides some indication of the precision of the estimate. The underlying principle is that an examinee, hypothetically, could take the same test a large number of times, assuming no recall of how the previous test items were answered. An ability estimate $\hat{\theta}$ would be obtained from each testing. The standard error is a measure of the variability of the values of $\hat{\theta}$ around the examinee's unknown parameter value θ. In the present case an estimated standard error can be computed using the equation given as

$$\text{SE}(\hat{\theta}) = \frac{1}{\sqrt{\sum_{j=1}^{J} a_j^2 P_j(\hat{\theta}) Q_j(\hat{\theta})}}. \tag{5.2}$$

It is of interest to note that the term under the square root sign is exactly the negative denominator term in Eq. (5.1). As a result, the estimated standard error can be obtained as a side product of estimating the examinee's ability.

In the example given above, it was

$$\text{SE}(\hat{\theta} = 0.32) = 1/\sqrt{0.6616} = 1.22944624.$$

Thus, the examinee's ability was not estimated very precisely as the standard error, 1.23, is very large. This is primarily due to the fact that only three items were used here and one would not expect a very good estimate. As will be shown in the next chapter, the standard error of an examinee's estimated ability plays an important role in item response theory.

There are two cases for which the maximum likelihood estimation procedure fails to yield a finite ability estimate. First, when an examinee answers none of the items correctly, the corresponding ability estimate is negative infinity. Second, when an examinee answers all the items in the test correctly, the corresponding ability estimate is positive infinity. In both of these cases it is impossible to obtain an ability estimate for the examinee (the computer literally cannot compute a number as big as infinity). Consequently, the computer programs used to estimate ability must protect themselves against these two conditions. When they detect either a test score of zero or a perfect test score, they will eliminate the examinee from further analysis and set the estimated ability to some symbol such as ****** to indicate what has happened. Sometimes a fixed set of finite numbers (e.g., $-\log 2J$ for a test score of zero and $\log 2J$ for a perfect score) can be arbitrary assigned to these cases.

5.3 Item Invariance of an Examinee's Ability Estimate

Another basic principle of item response theory is that the examinee's ability is invariant with respect to the items used to determine it. This principle rests upon two conditions: first, all the items measure the same underlying latent trait; second, the values of all the item parameters are in a common metric. To illustrate this principle, assume that an examinee has an ability score of zero, which places him at the middle of the ability scale. Now, if a set of ten items having an average item difficulty of -2.0 were administered to this examinee, the item responses can be used to estimate the examinee's ability, yielding $\hat{\theta}_1$ for this test. Then if a second set of ten items having an average item difficulty of $+1.0$ were administered to this examinee, these item responses can be used to estimate the examinee's ability, yielding $\hat{\theta}_2$ for this second test. Under the item-invariance principle, $\hat{\theta}_1 = \hat{\theta}_2$; that is, the two sets of items should yield the same ability estimate, within sampling variation, for the examinee. In addition, there is no requirement that the item discrimination parameters be the same for the two sets of items. This principle is just a reflection of the fact that the item characteristic curve spans the whole ability scale. Just as any subrange of the ability scale can be used in the estimation of item parameters, the corresponding segments of several item characteristic curves can be used to estimate an examinee's ability. Items with a high average item difficulty will have a point on their item characteristic curves that corresponds to the ability of interest. Similarly, items with a low average item difficulty will have a point on their item characteristic curves that corresponds to the ability of interest. Consequently, either set of items can be used to estimate the ability of examinees at that point. In each set, a different part of the item characteristic curve is involved, but that is acceptable.

The practical implication of this principle is that a test located anywhere along the ability scale can be used to estimate an examinee's ability. For example, an examinee could take a test that is "easy" or a test that is "hard" and obtain, on the average, the same estimated ability. This is in sharp contrast to classical test theory where such an examinee would get a high score on the easy test, a low score on the hard test, and there is no way of ascertaining the examinee's underlying ability. Under item response theory, the examinee's ability is fixed and invariant with respect to the items used to measure it. A word of warning is in order with respect to the meaning of the word "fixed." An examinee's ability is fixed only in the sense that it has a particular value in a given context. For example, if an examinee took the same test several times and it could be assumed he or she would not remember the items or the responses from testing to testing, the examinee's ability would be fixed. However, if the examinee received remedial instruction between testings or if there were carryover effects, the examinee's underlying ability level would be different for each testing. Thus, the examinee's underlying ability level is not immutable. There are a number of applications of item response theory that depend upon an examinee's ability level changing as a function of changes in the educational context.

The item invariance of an examinee's ability and the group invariance of an item's parameters are two facets of what is referred to, generically, as the invariance principle of item response theory. This principle is the basis for a number of practical applications of the theory.

5.4 Computer Session

This session has three purposes that result in apparently similar outcomes which are actually different in their conceptual basis. The first is to show how to obtain the estimate of the ability parameter under the two-parameter model. A general function to estimate the ability parameter for all three item characteristic curve models is also presented in conjunction with the first purpose under the assumption that the values of item parameters are known.

The second purpose is to illustrate how an examinee's estimated ability varies when the same test is taken a number of times. A test consisting of a few items with known item parameters will be established, the value of the examinee's ability parameter will be set, and the computer will generate the examinee's item responses. These will be used in Eq. (5.1) to estimate the examinee's ability. The computer will then generate a new set of item responses to these same items and another ability estimate is to be obtained. After several estimates are obtained, the mean and standard deviation of the estimates will be computed and compared to their theoretical values. The intent is to allow you to develop a sense of how ability estimates for a single examinee are distributed under repeated use of the same test.

The third purpose is to illustrate the item invariance of an examinee's ability. A small test will be established through the values of its item parameters, the examinee's ability will be set, and the computer will generate the examinee's item

5.4 Computer Session

responses. These will be used in Eq. (5.1) to obtain an ability estimate for the examinee. Then a new test with a set of different item parameter values will be established, item responses for the same examinee generated, and another ability estimate obtained. This process will be repeated for several different tests, resulting in a set of ability estimates. If the invariance principle holds, all the estimates should be clustered around the value of the examinee's ability parameter.

5.4.1 Procedures for an Example Case

This example will illustrate how to obtain an ability estimate for an examinee who responded to a test with three items. It is assumed that the item parameters under the two-parameter model are known. The followings are the R command lines to obtain the ability estimate and the standard error:

```
> u <- c(1, 0, 1)
> b <- c(-1.0, 0.0, 1.0)
> a <- c(1.0, 1.2, 0.8)
> th <- 1.0
> J <- length(b)
> S <- 10
> ccrit <- 0.001
> for (s in 1:S) {
    sumnum <- 0.0
    sumdem <- 0.0
    for (j in 1:J) {
      phat <- 1 / (1 + exp(-a[j] * (th - b[j])))
      sumnum <- sumnum + a[j] * (u[j] - phat)
      sumdem <- sumdem - a[j]**2 * phat * (1.0 - phat)
    }
    delta <- sumnum / sumdem
    th <- th - delta
    cat(paste("th=", th, "\n")); flush.console()
    if (abs(delta) < ccrit | s == S) {
      se <- 1 / sqrt(-sumdem)
      cat(paste("se=", se, "\n")); flush.console()
      break
    }
  }
> th
> se
```

By pressing the enter key in the end of each line, the computer will perform iterations to obtain the ability estimate and the accompanying standard error for an examinee with the given item responses using the known item parameters under the two-parameter model. The final values of the ability estimate and the standard error can be printed out with the last two command lines after finishing the loop.

The first line contains the examinee's response vector. The next two command lines are used to set up the known item parameters for the three items. The fourth line is used to set up an initial value of the $\hat{\theta}$, th (i.e., theta hat). To match the initial value to the one used in the illustration in the beginning of this chapter, the value of 1.0 was used here. The fifth line defines the total number of items J to the length of the vector of the item difficulty parameters. The sixth line sets up the maximum number of iterations to perform (i.e., 10 in this example). The seventh line defines the value of the convergence criterion, ccrit, for the iteration. If the absolute value of the $\Delta\hat{\theta}_s$ term is smaller than the specified convergence criterion value (i.e., 0.001 in the above command line), then the iterative process will be terminated.

The for loop in the next line performs the iterations until the change in the absolute value of the adjustment term reaches to a value smaller than that of the convergence criterion. After initializing the sum of the numerator and the sum of the denominator of the adjustment term in Eq. (5.1), the inner for loop will obtain the values of $P_j(\hat{\theta}_s)$, $a_j[u_j - P_j(\hat{\theta}_s)]$, and $a_j^2 P_j(\hat{\theta}_s) Q_j(\hat{\theta}_s)$ for each item. After obtaining these values and finishing the inner for loop, the lines in the subsequent outer for loop will print out, using the R functions cat and paste, the improved provisional value of $\hat{\theta}_{s+1}$ in the R console window. When the convergence criterion has been met or the iteration reaches the maximum number of iterations, then the standard error from Eq. (5.2) will be calculated and printed into the R console window.

5.4.2 An R Function to Estimate Ability

It is possible to create an R function to estimate the ability parameter and to obtain the standard error of the estimate given the item characteristic curve model, the response vector, and the set of known item parameters. Consider the following function named ability:

```
>   ability <- function(mdl, u, b, a, c) {
      J <- length(b)
      if (mdl == 1 | mdl == 2 | missing(c)) {
        c <- rep(0, J)
      }
      if (mdl == 1 | missing(a)) { a <- rep(1, J) }
      x <- sum(u)
      if (x == 0) {
        th <- -log(2 * J)
      }
      if (x == J) {
        th <- log(2 * J)
      }
      if (x == 0 | x == J) {
```

5.4 Computer Session

```
      sumdem <- 0.0
      for (j in 1:J) {
        pstar <- 1 / (1 + exp(-a[j] * (th - b[j])))
        phat <- c[j] + (1.0 - c[j]) * pstar
        sumdem <- sumdem - a[j]**2 * phat * (1.0 - phat) *
          (pstar / phat)**2
      }
      se <- 1 / sqrt(-sumdem)
    }
    if (x != 0 & x != J) {
      th <- log(x / (J - x))
      S <- 10
      ccrit <- 0.001
      for (s in 1:S) {
        sumnum <- 0.0
        sumdem <- 0.0
        for (j in 1:J) {
          pstar <- 1 / (1 + exp(-a[j] * (th - b[j])))
          phat <- c[j] + (1.0 - c[j]) * pstar
          sumnum <- sumnum + a[j] * (u[j] - phat) *
            (pstar / phat)
          sumdem <- sumdem - a[j]**2 * phat * (1.0 - phat) *
            (pstar / phat)**2
        }
        delta <- sumnum / sumdem
        th <- th - delta
        if (abs(delta) < ccrit | s == S) {
          se <- 1 / sqrt(-sumdem)
          break
        }
      }
    }
    cat(paste("th=", th, "\n")); flush.console()
    cat(paste("se=", se, "\n")); flush.console()
    thse <- c(th, se)
    return(thse)
  }
```

The input to the function ability first depends on the item characteristic curve model, mdl, where the three integers, 1, 2, and 3, are acceptable. The response vector, u, that is the same length as each of the item parameter vectors is also required. The set of item parameters dependent upon the model selected is also required in the default order of b, a, and c. Using the required input variables as the arguments of the function, the function obtains and prints out the value of the ability estimate (th= $\hat{\theta}$) and the standard error of the ability estimate (se= SE($\hat{\theta}$)). Ultimately the function returns a vector of size two (i.e., thse–theta hat and standard error) that contains the ability estimate and the standard error of the ability estimate.

There are several things that should be explicated. First, unlike the earlier example the initial value of the ability estimate in the function `ability` is $\log(x/(J-x))$, where x is the summed, raw score of the examinee and J is the total number of items. Second, before setting up the initial ability estimate of the log odds ratio type, finite ability estimates, $-\log(2J)$ and $\log(2J)$, will be arbitrarily assigned to the respective values of $x = 0$ and $x = J$. This may not be a universally viable solution to the cases of a raw score of 0 and a perfect raw score of J; $-\log(2J-1)$ and $\log(2J-1)$ can also be used. As mentioned earlier, many computer programs will simply print out a statement that finite ability estimates cannot be obtained for these cases. With the finite estimates, the standard errors of the estimate can be obtained for $x = 0$ and $x = J$. Third, the standard error of the estimate is not entirely based on Eq. (5.2) because the three-parameter model requires some modification. Last, the numerator term in Eq. (5.1) is also modified under the three-parameter model.

For the example data, the ability estimate and the standard error can be obtained after defining the function `ability`, and then type in the following command lines to the R console window:

```
> u <- c(1, 0, 1)
> b <- c(-1.0, 0.0, 1.0)
> a <- c(1.0, 1.2, 0.8)
> ability(2, u, b, a)
```

You may notice that the values of the ability estimate and the standard error from the function `ability` are not exactly the same as those from the earlier results. The different initial values caused the trivial differences.

If you want to make the output from the function `ability` as a variable, the last line can be replaced with:

```
> thse <- ability(2, u, b, a)
> thse
```

The values contained in the vector `thse` within the function `ability` will be returned, that is, newly created as an R variable with the same name, after executing the function. The values are displayed by typing in the variable name. You may use a different name; so instead of using the above command line you may use:

```
> theta.se <- ability(2, u, b, a)
> theta.se
```

5.4.3 Procedure for Investigating the Sampling Variability of Estimated Ability

This example case is to illustrate the sampling variability of a given examinee's estimated ability when the same test is administered several times. The model, the

5.4 Computer Session

set of item parameters, and the ability parameter are assumed to be known. The ability parameter can be randomly generated using an R function instead of being arbitrarily specified. The R command line

```
> theta <- rnorm(1, 0, 1)
```

can be used for which the arguments in the R function `rnorm` (i.e., randomly generate normal deviates) designate the number of cases, the mean, and the standard deviation of the normal distribution. Then the item response vector will be randomly generated based on the parameters. The generated item response vector will be used to obtain the ability estimate and the standard error. The process will be replicated several times. Theoretically, the ability estimates from the replication should be very close to the ability parameter that was used to generate the data. The standard deviation of the ability estimates should be very similar to the theoretical standard error based on the ability parameter. Also the average of the standard errors of the estimates should be very similar to the theoretical standard error. The function `ability` will be employed for each replication, and hence the values of item parameters are assumed to be known and used as arguments.

The R command lines are as follows:

```
> mdl <- 2
> theta <- 0.5
> b <- c(-0.5, -0.25, 0.0, 0.25, 0.5)
> a <- c(1.0, 1.5, 0.7, 0.6, 1.8)
> J <- length(b)
> if (mdl == 1 | mdl == 2) { c <- rep(0, J) }
> if (mdl == 1) { a <- rep(1, J) }
> sumdemt <- 0.0
> for (j in 1:J) {
    Pstar <- 1 / (1 + exp(-a[j] * (theta - b[j])))
    P <- c[j] + (1 - c[j]) * Pstar
    sumdemt <- sumdemt - a[j]**2 * P * (1.0 - P) *
       (Pstar / P)**2
  }
> set <- 1 / sqrt(-sumdemt)
> R <- 10
> thr <- rep(0, R)
> ser <- rep(0, R)
> for (r in 1:R) {
    u <- rep(0, J)
    for (j in 1:J) {
      P <- c[j] + (1 - c[j]) /
         (1 + exp(-a[j] * (theta - b[j])))
      u[j] <- rbinom(1, 1, P)
    }
    thse <- ability(mdl, u, b, a, c)
    thr[r] <- thse[1]
```

```
          ser[r] <- thse[2]
     }
> theta
> set
> thr
> mean(thr)
> sd(thr)
> ser
> mean(ser)
```

The last seven lines will generate the main results. The ability parameter, 0.5, that we specified in the earlier input line will be printed out in the R console window from the `theta` line. The theoretical standard error from the parameter value as if the parameter is hypothesized to be an estimate will be printed by the next line `set` (i.e., standard error–theoretical). Based on the number of replications specified, the vector that contains the ability estimates from the replications will be printed out by `thr` (i.e., theta hat from the replication). The average value of the `thr` will be obtained using the R function `mean`. The observed standard error will be obtained as the standard deviation of the replicated ability estimates using the R function `sd`. The vector that contains the standard errors of the estimates from the replications will be printed out by `ser` (i.e., standard error from the replication). The average value of the standard errors will be calculated in the last line.

Because the function `ability` contains the command lines to print out the ability estimate and the standard error to the R console window in each iteration, the values of `thr` and `ser` can also be found in the beginning of the output lines. If these values from the replications are not wanted, the two lines in the function `ability` should be commented out with a symbol # or possibly deleted from the function before executing it:

```
#    cat(paste("th=", th, "\n")); flush.console()
#    cat(paste("se=", se, "\n")); flush.console()
```

Parts of the output in the R console window from one example run are as follows:

```
...
> theta
[1] 0.5
> set
[1] 0.7829669
> thr
 [1]  0.2622246 -0.2226930 -0.2874176  1.8696771  0.3215108
 [6] -0.8031985  0.8995979  1.7040941  0.4412501  0.8995979
> mean(thr)
[1] 0.5084643
> sd(thr)
[1] 0.8548544
> ser
 [1]  0.7693594 0.7999961 0.8092524 1.3458059 0.7707814
```

5.4 Computer Session

```
 [6] 0.9231352 0.8577812 1.2321971 0.7776090 0.8577812
> mean(ser)
[1] 0.9143699
>
```

Note that the ability parameter, 0.5, is very close to the mean of the ability estimates, 0.51. The theoretical standard error, 0.78, is somewhat smaller than the observed standard error obtained from the ability estimates (i.e., 0.85) or the average value of the standard errors from the ten replications (i.e., 0.91). The number in the brackets in the R output indicates the order of the elements in a variable or vector. Such a number is useful especially if the variable is a vector of some length (e.g., `thr`). For example, the sixth value of the ability estimates from the replications is printed out the right side of [6]. Depending upon the size of the R console window on the computer screen, the number within the brackets in the second line of the output for a variable display will be changed. The observed standard error of the ability estimates should approximate the theoretical value. However, with such a small number of items and replications, the results will probably deviate somewhat from their theoretical values.

Although in the example run the mean of the ability estimates is very close to the ability parameter, each ability estimate does not seem to be very close to the ability parameter. In addition, other runs due to the random data generation may yield the mean of the ability estimates that is not really similar to the parameter value. Also standard errors from the iterations might show somewhat different sizes. When the number of items is as large as the usual number of items in practical testing situations, all the estimates will be similar and possibly very close to the ability parameter value. A large number of items also yields values of standard errors to be smaller and similar.

5.4.4 Procedure for Investigating the Item Invariance of an Examinee's Ability

In this example a given examinee will be administered a number of different tests. The intent is to illustrate that the estimated abilities should cluster about the value of the examinee's ability parameter. The function `ability` will be employed in each replication to obtain the ability estimate and the standard error.

The model, the value of the ability parameter, and the number of items in a test are assumed to be known. The computer will generate the random values of the item parameters. Using these parameters and the specified item characteristic curve model, the item response vector will be randomly generated. The examinee's ability will be estimated and shown in the R console window accompanied by its standard error. The process will be replicated several times. The maximum number of iterations is specified by the value of *R*. Theoretically, the ability estimates from replications should be very close to the ability parameter that was used to generate the data. The average of the ability estimates can be obtained and compared to the value of the ability parameter. There should not be a large amount of scatter in the estimates. Again, due to the small number of items and the limited number of ability estimates from the replications used, the item invariance of the ability estimate may not be readily apparent.

The R command lines for an example run are as follows:

```
mdl <- 2
theta <- 0.5
J <- 5
R <- 10
thr <- rep(0, R)
ser <- ref(0, R)
for (r in 1:R) {
   b <- round(runif(J,-3,3), 2)
   a <- round(runif(J,0.2,2.8), 2)
   c <- round(runif(J,0,.35), 2)
   if (mdl == 1 | mdl == 2) { c <- rep(0, J) }
   if (mdl == 1) { a <- rep(1, J) }
   u <- rep(0, J)
   for (j in 1:J) {
      P <- c[j] + (1 - c[j]) /
         (1 + exp(-a[j] * (theta - b[j])))
      u[j] <- rbinom(1, 1, P)
   }
   thse <- ability(mdl, u, b, a, c)
   thr[r] <- thse[1]
   ser[r] <- thse[2]
}
theta
thr
mean(thr)
```

The last three lines will generate the main results. The ability parameter, 0.5, that we specified in the earlier input line will be printed out in the R console window from the theta line. Based on the number of replications specified, the vector that contains the ability estimates from the replications will be printed out by thr (i.e., theta hat from the replication). The average value of the thr will be obtained using the R function mean.

Because the function ability contains the command lines to print out the ability estimate and the standard error to the R console window in each iteration, the values of thr and ser can also be found in the beginning of the output lines. Again, if these values from the replications are not wanted, the two lines in the function ability should be commented out with a symbol # or possibly deleted from the function before executing it.

5.4 Computer Session

Parts of the output in the R console window from one example run are as follows:

```
...
> theta
[1] 0.5
> thr
[1]  1.42668348 -1.26110250 -0.04506560  1.10788283
[5] -0.03994666  2.30258509 -0.20780057  0.59002658
[9]  0.30427844  0.67874115
> mean(thr)
[1] 0.4856282
>
```

Note that the ability parameter, 0.5, is close to the mean of the ability estimates, 0.49, for this particular example run. However, with such a small number of items and the limited number of replications, the results will probably deviate somewhat from their theoretical values. Each ability estimate does not seem to be very close to the ability parameter. Other runs due to the random data generation may yield the mean of the ability estimates that is not really similar to the parameter value. When the number of items is as large as the usual number of items in practical testing situations, all the estimates will be similar and very close to the ability parameter value. Also high quality item parameters obtained from narrow ranges of the uniform distribution (e.g., b values matched with the ability parameter, high a values, and near-zero c values) may yield the ability estimate that is close to the ability parameter.

The number of items can be increased by changing the value of J. The item parameters can also be generated via some other distributions, for example:

```
b <- rnorm(J, 0, 2)
a <- rlnorm(J, 0, 0.5)
c <- rbeta(J, 41, 161)
```

In the above specification, values of the item difficulty parameters will be randomly sampled from the normal distribution with parameters of mean 0 and standard deviation 2; values of the item discrimination parameters will be randomly sampled from the lognormal distribution with parameters of mean 0 and standard deviation 0.5; values of the guessing parameters will be randomly sampled from the beta distribution with parameters of alpha 41 and beta 161 (cf. the mode is $(41-1)/(41+161-2) = 0.2$; the mean is $41/(41+161) = 0.202970297$).

Better quality item parameters can be generated by changing the arguments in the respective distributions. For example, assuming that the ability parameter is located in the middle of the ability scale at 0, we may specify:

```
b <- rnorm(J, 0, 1)
a <- rlnorm(J, 0, 0.25)
c <- rbeta(J, 401, 1601)
```

If only two decimal places are needed for these parameters, then we may use:

```
b <- round(rnorm(J,0,1), 2)
a <- round(rlnorm(J,0,0.25), 2)
c <- round(rbeta(J,401,1601), 2)
```

5.5 Exercises

For the following exercises, it is assumed that you have defined the function `ability` by typing it in the R console window.

5.5.1 Sampling Variability of Estimated Ability

1. An example case earlier will be performed using the item characteristic curve model of your choice. The two-parameter model may be a good starting model.
    (a) Use the number of items $J = 5$ and the number of replications $R = 10$.
    (b) Set up the ability parameter to 0.5. Note that you may use a randomly sampled value instead.
    (c) Use the item characteristic curve model that you want to employ.
    (d) Since you know the value of the examinee's ability parameter, choose values of the item difficulty parameters randomly from a uniform distribution with a range from −3 to 3. If the two-parameter model is used, choose values of the item discrimination parameters randomly from a uniform distribution with a range from 0.2 to 2.8. For the three-parameter model, try to use values of the guessing parameters randomly selected from a uniform distribution with a range from 0 to 0.35. You may use the following R command lines:

    ```
    b <- runif(J, -3, 3)
    a <- runif(J, 0.2, 2.8)
    c <- runif(J, 0, .35)
    ```

    Almost equivalently, you may use the following command lines, assuming that you want to use only two decimal places for the values of item parameters:

    ```
    b <- round(runif(J,-3,3), 2)
    a <- round(runif(J,0.2,2.8), 2)
    c <- round(runif(J,0,.35), 2)
    ```

    (e) Obtain the ability estimates and standard errors as well as relevant statistics.

5.5 Exercises

When the summary results are obtained on the R console window, try to write down values of the ability estimates on a piece of paper so you can use them in the next exercise.

2. The intent of this exercise is to see if you can improve upon the previous estimate of the examinee's ability parameter by proper selection of the test's item parameters.

   (a) Use the number of items $J = 5$ and the number of replications $R = 10$.
   (b) Set up the ability parameter to one of the values you have obtained from the previous exercise (i.e., a number may be close to the original ability parameter 0.5).
   (c) Use the item characteristic curve model that you employed in the previous exercise.
   (d) Since you know the value of the examinee's ability parameter, choose values of the item difficulty parameters that are close to this ability parameter value and use large values of item discrimination parameters. For the three-parameter model, try to use the guessing parameter values a bit smaller than 0.2.
   (e) Obtain the ability estimates and standard errors as well as relevant statistics.

   If you chose the item parameter values wisely, the mean of the ability estimates should have been close to the value of the examinee's ability parameter. The observed standard error should have also approximated the theoretical value. If such was not the case, first, think about some reasons for the lack of match. You need to keep in mind that the obtained results are subject to considerable sampling variability due to the small numbers of items being used (increasing J to 10 or a higher number, say 30, will help) and the limited number of replications used.

3. Experiment with different types of models and item parameter values to see if you can determine what influences the distribution of the estimated abilities.

5.5.2 Item Invariance of an Examinee's Ability

1. The intent of this exercise is to enable you to experiment with the item sets used to illustrate the item invariance of the ability estimates. Rather than letting the computer set the values of the item parameters, you can choose your own values.

   (a) Select the item characteristic curve model of your choice (e.g., the three-parameter model).
   (b) Set the ability parameter to a value of your choice (e.g., $\theta = 0.5$).
   (c) Use the number of items $J = 5$.
   (d) Select the number of replications, say $R = 4$. You may set up and initialize the variables that will hold the ability estimates and the standard errors.

(e) Set up the item parameter values of your own. For example, for the three-parameter logistic model you may use such command lines as:

```
b <- c(-0.5, -0.25, 0.0, 0.25, 0.5)
a <- c(1.0, 1.5, 0.7, 0.6, 1.8)
c <- c(.1, .2, .1, .15, .2)
```

(f) Use R command lines to create an examinee's item response vector that contains the randomly generated responses of $J = 5$ items based on the ability parameter and the item parameters.

(g) Use the function ability to obtain the returned vector that contains the values of the ability estimate and the standard error.

(h) Write down the ability estimate and the standard error or try to save the values onto the variables initialized in step e.

(i) Repeat steps e through h $R = 4$ times, but using a different set of item parameters in each replication.

(j) Obtain the average value of the ability estimates from the replications and compare the value with the ability parameter.

Theoretically, the average value of the estimates should be close to the value of the examinee's ability parameter. There should not be a large amount of scatter in the estimates. Again due to the small number of items and the limited number of estimates used, the item invariance of the ability estimate may not be readily apparent.

2. To make things easy for you, you can let the computer generate not only the ability parameter but also the sets of item parameter values. Repeat the procedures for Exercise 1. Now the computer will do the tedious job of setting the item parameters. You may experiment with a large number of items J as well as a large number of replications R.

5.6 Things to Notice

1. Distribution of estimated ability.

   (a) The average value of the estimates is reasonably close to the value of the ability parameter for the examinee set by the computer program.
   (b) When the item difficulties are at or near the examinee's ability parameter value, the mean of the estimated abilities will be close to that ability value.
   (c) The standard error of the estimates can be quite large when the items are not located near the ability of the examinee. However, the theoretical values of the standard errors are also quite large and the obtained standard errors approximate these values.
   (d) When the values of the item discrimination parameters are large, the standard error of the ability estimates is small. When the item discrimination parameters are small, the standard error of the ability estimates is large.

5.6 Things to Notice

(e) You may have noticed that the number of iterations needed to obtain the ability estimate varies widely. The closer the item difficulty parameters were to the examinee's ability, the fewer the number of iterations needed.

(f) The optimum set of items for estimating an examinee's ability would have all its item difficulty parameters equal to the examinee's ability parameter and have items with large values for the item discrimination parameters.

2. Item invariance of the examinee's ability.

   (a) The different sets of items can yield values of estimated ability that are near the examinee's actual ability level.

   (b) The mean value of these estimates generally is a close approximation of the examinee's ability parameter. If one used many tests, each having a large number of items, the mean estimated ability would equal the examinee's ability parameter. In addition, these estimates would be very tightly clustered around the parameter value. In such a situation, it would be very clear that the item invariance principle holds.

3. Overall observation.

   (a) The computer session and exercises have mainly dealt with two facets of estimating an examinee's ability that are conceptually distinct but look similar in certain respects. The first set of examples focused upon the variability of the ability estimates about the value of the examinee's ability parameter. This will serve as the basis for the next chapter, which deals with how well a test estimates ability over the whole ability scale. The second set of exercises focused upon the item invariance of an examinee's estimated ability. This will serve as part of the basis for Chap. 7 dealing with test calibration.

   (b) The reader should keep in mind that an ability estimate is just another type of test score, but it is interpreted within the context of item response theory. Consequently, such ability estimates can be used to compute summary statistics for groups of examinees and other indices of interest.

4. A final comment: In Chap. 1, the concept of a latent trait was introduced. An integral part of item response theory is that an examinee can be positioned on the scale representing this latent trait. Thus, in theory, each examinee has an ability score (parameter value) that locates that person on the scale. However, in the real world we cannot obtain the value of the examinee's ability parameter. The best one can do is obtain an estimate of it. In the computer session for this chapter, it was assumed that the ability parameter was known. We could either assign or generate the value of the examinee's ability parameter. The specified set of the ability parameter and the item parameters enabled the R program to generate the item response vectors used to obtain the ability estimates and hence to illustrate the theory.

Chapter 6
The Information Function

6.1 Introduction

When you speak of having information, it implies that you know something about a particular object or topic. In statistics and psychometrics, the term information conveys a similar, but somewhat more technical, meaning. The statistical meaning of information is credited to Sir R.A. Fisher, who defined information as the reciprocal of the variance with which a parameter could be estimated. Thus, if you could estimate a parameter with precision (i.e., smaller variability), you would know more about the value of the parameter than if you had estimated it with less precision (i.e., larger variability). Statistically, the magnitude of precision with which a parameter is estimated is inversely related to the size of the variability of the estimates around the value of the parameter. The variance of the estimators is denoted by σ^2. The amount of information, denoted by I, then is given by the formula

$$I = \frac{1}{\sigma^2}. \tag{6.1}$$

In item response theory, our interest is in estimating the value of the ability parameter for an examinee. The ability parameter is denoted by θ and $\hat{\theta}$ is an estimator of θ. In the previous chapter, the standard deviation of the ability estimates about the examinee's ability parameter was computed. If this term is squared, it becomes a variance. The inverted variance is a measure of the precision with which a given ability level can be estimated. From Eq. (6.1), the amount of information at a given ability level is the reciprocal of this variance. If the amount of information is large, it means that an examinee whose true ability is at that level can be estimated with precision; that is, all the estimates will be reasonably close to the true value. If the amount of information is small, it means that the ability cannot be estimated with precision and the estimates will be widely scattered about the true ability. Using the appropriate formula, the amount of information can be computed for each ability

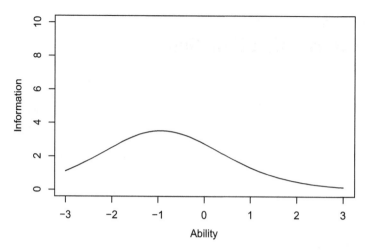

Fig. 6.1 An information function

level on the ability scale from negative infinity to positive infinity. Because ability is a continuous variable, information will also be a continuous variable. If the amount of information is plotted against ability, the result is a graph of the information function such as that shown in Fig. 6.1.

Inspection of Fig. 6.1 shows that the amount of information has a maximum at an ability level of -1.0 and is about 3 for the ability range of $-2.0 \leq \theta \leq 0.0$. Within this range, ability is estimated with some precision. Outside this range the amount of information decreases rapidly and the corresponding ability levels are not estimated very well. Thus, the information function tells us how well each ability level is being estimated. It is important for the reader to recognize that the information function does not depend upon the distribution of examinees over the ability scale. In this regard, it is like the item characteristic curve and the test characteristic curve. In a general-purpose test, the ideal information function would be a horizontal line at some large value of I and all ability levels would be estimated with the same precision. Unfortunately, such an information function is hard to achieve. The typical information function looks somewhat like that shown in Fig. 6.1 and different ability levels are estimated with differing degrees of precision. This becomes of considerable importance to both the test constructor and the test consumer since it means that the precision with which an examinee's ability is estimated by a given test, depends upon where the examinee's ability is located on the ability scale.

6.2 Item Information Function

Since it depends upon the individual items composing a test, item response theory is what is known as an itemized theory. Under the theory, each item of the test measures the underlying latent trait. As a result the amount of information, based

6.3 Test Information Function

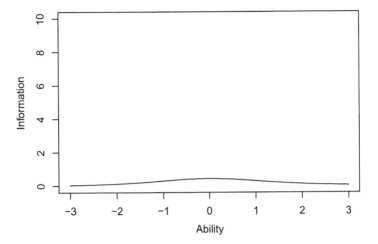

Fig. 6.2 An item information function

upon a single item, can be computed at any ability level and is denoted by $I_j(\theta)$ where j indexes the item. Because only a single item is involved, the amount of information at any point on the ability scale is going to be rather small. If the amount of item information is plotted against ability, the result is a graph of the item information function such as that shown in Fig. 6.2.

An item measures ability with greatest precision at the ability level corresponding to the item's difficulty parameter. The amount of item information decreases as the ability level departs from the item difficulty and approaches zero at the extremes of the ability scale.

6.3 Test Information Function

Since a test is used to estimate the ability of an examinee, the amount of information yielded by the test at any ability level can also be obtained. A test is a set of items; therefore, the test information at a given ability level is simply the sum of the item informations at that level. Consequently, the test information function is defined as

$$I(\theta) = \sum_{j=1}^{J} I_j(\theta), \tag{6.2}$$

where

$I(\theta)$ is the amount of test information at an ability level of θ,
$I_j(\theta)$ is the amount of information for item j at ability level θ, and
J is the number of items in the test.

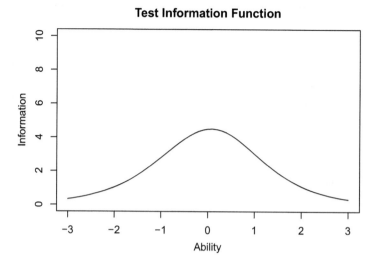

Fig. 6.3 A test information function

The general level of the test information function will be much higher than that for a single item information function. Thus, a test measures ability more precisely than does a single item. An important feature of the definition of test information given in Eq. (6.2) is that the more items in the test, the greater the amount of information. Thus, in general, longer tests will measure an examinee's ability with greater precision than will shorter tests. Plotting the amount of test information against ability yields a graph of the test information function such as that shown in Fig. 6.3 for a ten-item test.

The maximum value of the test information function in Fig. 6.3 is modest and, in this example, the amount of information decreases rather steadily as the ability level differs from that corresponding to the maximum. Thus, ability is estimated with some precision near the center of the ability scale. However, as the ability level approaches the extremes of the scale, the amount of test information decreases significantly.

The test information function is an extremely useful feature of item response theory. It basically tells you how well the test is doing in estimating ability over the whole range of ability scores. While the ideal test information function often may be a horizontal line, it may not be the best for a specific purpose. For example, if you were interested in constructing a test to award scholarships this ideal may not be optimal. In this situation, you would like to measure ability with considerable precision at ability levels near the ability used to separate those who will receive the scholarship from those who do not. The best test information function in this case would have a peak at the cutoff score. Other specialized uses of tests could require other forms of the test information function.

While an information function can be obtained for each item in a test, this is rarely done. The amount of information yielded by each item is rather small and we typically do not attempt to estimate an examinee's ability with a single item. Consequently, the amount of test information at an ability level and the test information function are of primary interest. Since the test information is obtained by summing the items information at a given ability level, the amount of information is defined at the item level. The mathematical definition of the amount of item information depends upon the particular item characteristic curve model employed. Therefore, it is necessary to examine these definitions under each model.

6.4 Definition of Item Information

6.4.1 Two-Parameter Item Characteristic Curve Model

Under the two-parameter model, the item information function is defined as

$$I_j(\theta) = a_j^2 P_j(\theta) Q_j(\theta), \qquad (6.3)$$

where

a_j is the item discrimination parameter for item j,
$P_j(\theta) = 1/[1 + \exp(-L_j)]$,
$L_j = a_j(\theta - b_j)$,
$Q_j(\theta) = 1 - P_j(\theta)$, and
θ is the ability level of interest.

To illustrate the use of Eq. (6.3), the amount of item information will be computed at seven ability levels for an item having parameter values of $b_j = 1.0$ and $a_j = 1.5$ (see Table 6.1).

Table 6.1 Calculation of item information under the two-parameter model, $b_j = 1.0$ and $a_j = 1.5$

θ	L_j	$\exp(-L_j)$	$P_j(\theta)$	$Q_j(\theta)$	$P_j Q_j$	a_j^2	$I_j(\theta)$
−3.0	−6.0	403.429	0.002	0.998	0.002	2.25	0.006
−2.0	−4.5	90.017	0.011	0.989	0.011	2.25	0.024
−1.0	−3.0	20.086	0.047	0.953	0.045	2.25	0.102
0.0	−1.5	4.482	0.182	0.818	0.149	2.25	0.336
1.0	0.0	1.000	0.500	0.500	0.250	2.25	0.563
2.0	1.5	0.223	0.818	0.182	0.149	2.25	0.336
3.0	3.0	0.050	0.953	0.047	0.045	2.25	0.102

Note: $P_j Q_j = P_j(\theta) Q_j(\theta)$

This item information function increases rather smoothly as ability increases and reaches a maximum value of 0.563 at an ability of 1.0. After this point it decreases. The obtained item information function is symmetrical about the value of the item's difficulty parameter. Such symmetry holds for all item information functions under the Rasch model and the two-parameter model. When only a single item is involved and the discrimination parameter has a moderate value, the magnitude of the amount of item information is quite small.

6.4.2 Rasch Item Characteristic Curve Model

Under the Rasch model, the item information is defined as

$$I_j(\theta) = P_j(\theta)Q_j(\theta). \tag{6.4}$$

This is exactly the same as that under the two-parameter model when the value of the item discrimination parameter is set to 1.0. To illustrate the use of Eq. (6.4), the amount of item information will be calculated for an item having an item difficulty parameter of 1.0 (see Table 6.2).

The general level of the amount of information yielded by this item is somewhat lower than that of the previous example. This is a reflection of the value of the item discrimination parameter being smaller than that of the previous item. Again, the item information function is symmetric about the value of the item difficulty parameter.

6.4.3 Three-Parameter Item Characteristic Curve Model

In Chap. 2, it was mentioned that the three-parameter model does not possess the nice mathematical properties of the logistic function. The loss of these properties

Table 6.2 Calculation of item information under the Rasch model, $b_j = 1.0$

θ	L_j	$\exp(-L_j)$	$P_j(\theta)$	$Q_j(\theta)$	P_jQ_j	a_j^2	$I_j(\theta)$
−3.0	−4.0	54.598	0.018	0.982	0.018	1.0	0.018
−2.0	−3.0	20.086	0.047	0.953	0.045	1.0	0.045
−1.0	−2.0	7.389	0.119	0.881	0.105	1.0	0.105
0.0	−1.0	2.718	0.269	0.731	0.197	1.0	0.197
1.0	0.0	1.000	0.500	0.500	0.250	1.0	0.250
2.0	1.0	0.368	0.731	0.269	0.197	1.0	0.197
3.0	2.0	0.135	0.881	0.119	0.105	1.0	0.105

Note: $P_jQ_j = P_j(\theta)Q_j(\theta)$

6.4 Definition of Item Information

becomes apparent in the complexity of the equation given below for the amount of item information under this model:

$$I_j(\theta) = a_j^2 \left[\frac{Q_j(\theta)}{P_j(\theta)}\right] \left[\frac{(P_j(\theta) - c_j)^2}{(1 - c_j)^2}\right], \tag{6.5}$$

where

$P_j(\theta) = c_j + (1 - c_j) \times 1/[1 + \exp(-L_j)]$,
$L_j = a_j(\theta - b_j)$, and
$Q_j = 1 - P_j(\theta)$.

To illustrate the use of these formulas, the computations will be shown for an item having parameter values of $b_j = 1.0$, $a_j = 1.5$, $c_j = 0.2$. The values of b_j and a_j are the same as those for the preceding two-parameter example. The computations will be performed in detail at an ability level of $\theta = 0.0$:

$L_j = 1.5(0.0 - 1.0) = -1.5$
$\exp(-L_j) = 4.482$
$1/[1 + \exp(-L_j)] = 0.182$
$P_j(\theta) = c_j + (1 - c_j) \times 1/[1 + \exp(-L)] = 0.2 + 0.8(0.182) = 0.346$
$Q_j(\theta) = 1 - 0.346 = 0.654$
$Q_j(\theta)/P_j(\theta) = 0.654/0.346 = 1.891$
$(P_j(\theta) - c_j)^2 = (0.346 - 0.2)^2 = (0.146)^2 = 0.021$
$(1 - c_j)^2 = (1 - 0.2)^2 = (0.8)^2 = 0.64$
$a_j^2 = (1.5)^2 = 2.25$

Then,

$$I_j(\theta) = I_j(0.0) = (2.25)(1.891)[(0.021)/(0.64)] = 0.142$$

Clearly, this is more complicated than the computations for the previous two models, which are in fact, logistic models. The amount of item information computations for this item at seven ability levels is shown in Table 6.3.

Table 6.3 Calculation of the amount of item information under the three-parameter model, $b_j = 1.0$, $a_j = 1.5$, $c_j = 0.2$

θ	L_j	$\exp(-L_j)$	$P_j(\theta)$	$Q_j(\theta)$	Q_j/P_j	$(P_j - c_j)^2$	$I_j(\theta)$
-3.0	-6.0	403.429	0.202	0.798	3.951	0.00000	0.00005
-2.0	-4.5	90.017	0.209	0.791	3.790	0.00008	0.00103
-1.0	-3.0	20.086	0.238	0.762	3.203	0.00144	0.01621
0.0	-1.5	4.482	0.346	0.654	1.891	0.02130	0.14157
1.0	0.0	1.000	0.600	0.400	0.667	0.16000	0.37500
2.0	1.5	0.223	0.854	0.146	0.171	0.42779	0.25699
3.0	3.0	0.050	0.962	0.038	0.039	0.58073	0.08052

Note: $Q_j/P_j = Q_j(\theta)/P_j(\theta)$ and $(P_j - c_j)^2 = (P_j(\theta) - c_j)^2$

The shape of this information function is very similar to that for the preceding two parameter example in which $b_j = 1.0$ and $a_j = 1.5$. However, the general level of the values for the amount of information is lower. For example, at an ability level of $\theta = 0.0$, the item information was 0.142 under the three-parameter model and 0.336 under the two-parameter model having the same values of b_j and a_j. In addition, the maximum of the information function did not occur at an ability level corresponding to the value of the difficulty parameter. The maximum occurred at an ability level slightly higher than the value of b_j. Because of the presence of the terms $(1 - c_j)$ and $(P_j(\theta) - c_j)$ in Eq. (6.5), the amount of information under the three-parameter model will be less than under the two-parameter model having the same values of b_j and a_j. When they share common values of a_j and b_j, the information functions will be the same when $c_j = 0$. When $c_j > 0$, the three-parameter model will always yield less information. Thus, the item information function under the two-parameter model defines the upper bound for the amount of information under the three-parameter model. This is reasonable, as getting the item correct by guessing should not enhance the precision with which an ability level is estimated.

6.5 Computing a Test Information Function

Equation (6.2) defined the test information as the sum of the amounts of item information at a given ability level. Now that the procedures for calculating the amount of item information have been shown for the three item characteristic curve models, the test information function for a test can be computed. To illustrate this process, a five-item test will be used. The item parameters under the two-parameter model are as follows:

$$b_1 = -1.0 \quad b_2 = -0.5 \quad b_3 = 0.0 \quad b_4 = 0.5 \quad b_5 = 1.0$$
$$a_1 = 2.0 \quad a_2 = 1.5 \quad a_3 = 1.5 \quad a_4 = 1.5 \quad a_5 = 2.0$$

The amount of item information and the test information may be computed for the same seven ability levels used in the previous examples (see Table 6.4).

Table 6.4 Calculations for a test information function based upon five items

θ	Item information					Test information
	1	2	3	4	5	
−3.0	0.071	0.051	0.024	0.012	0.001	0.159
−2.0	0.420	0.194	0.102	0.051	0.010	0.776
−1.0	1.000	0.490	0.336	0.194	0.071	2.091
0.0	0.420	0.490	0.563	0.490	0.420	2.383
1.0	0.071	0.194	0.336	0.490	1.000	2.091
2.0	0.010	0.051	0.102	0.194	0.420	0.776
3.0	0.001	0.012	0.024	0.051	0.071	0.159

6.6 Interpreting the Test Information Function

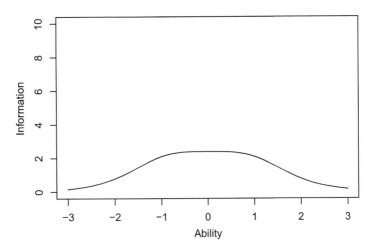

Fig. 6.4 Test information function for the five items of Table 6.4

Each of the item information functions was symmetric about the value of each item difficulty parameter. The five item discrimination parameters had a symmetrical distribution around a value of 1.5. The five item difficulty parameters had a symmetrical distribution about an ability level of zero. Because of this, the test information function also was symmetric about an ability of zero. The graph of this test information function is shown in Fig. 6.4.

The graph of the test information function shows that the amount of information was relatively flat over the range $\theta = -1$ to $\theta = +1$; outside of this range the amount of information decreased rather rapidly. However, in Table 6.4 the values of the test information varied over the whole ability scale. The apparent flat section of the plotted test information function in the graph is due to the coarseness and scattering of the item information functions.

6.6 Interpreting the Test Information Function

While the shape of the desired test information function depends upon the purpose for which a test is designed, some general interpretations can be made. A test information function that is peaked at some point on the ability scale measures ability with unequal precision along the ability scale. Such a test would be best for estimating the ability of examinees whose abilities fall near the peak of the test information function. In some tests, the test information function is rather flat over some region of the ability scale. Such tests estimate some range of ability scores with nearly equal precision and outside this range with less precision. Thus, the test would be a desirable one for those examinees whose ability falls in the given range.

When interpreting a test information function, it is important to keep in mind the reciprocal relationship between the amount of information and the variability of the ability estimates. To translate the amount of information into a standard error of estimate, one need only take the reciprocal of the square root of the amount of test information:

$$\text{SE}(\theta) = \frac{1}{\sqrt{I(\theta)}} \tag{6.6}$$

For example, in Fig. 6.4, the maximum amount of test information was 2.383 at an ability level of 0.0. This translates into a standard error of 0.648, which means roughly that 68% of the estimates of this ability level fall between -0.648 and $+0.648$ (i.e., the 68% confidence interval from $0.0 \pm 1 \times 0; 648$). Thus, this ability level is estimated with a modest amount of precision.

6.7 Computer Session

The purpose of this computer session is to enable you to develop a sense of how the form of the test information function depends upon the parameters of the items constituting the test. You will establish the parameter values for the items in a small test and then the computer will display the test information function on the screen. You can try different item characteristic curve models to determine how the choice of model affects the shape of the test information function. Under each model, different mixes of item parameter values can be used and the resultant test information function obtained. You should reach the point where you can predict the form of the test information function from the values of the item parameters.

6.7.1 Procedures for an Example Case

A graph of a test information function will be obtained using the R command lines. The number of items in the test is set to $J = 10$, and the item characteristic curve model is the two-parameter model. The calculation of information requires the item parameters are known. The values of the item parameters are as follows:

$$\begin{aligned}
b_1 &= -0.4 & a_1 &= 1.0 \\
b_2 &= -0.3 & a_2 &= 1.5 \\
b_3 &= -0.2 & a_3 &= 1.2 \\
b_4 &= -0.1 & a_4 &= 1.3
\end{aligned}$$

6.7 Computer Session

$$b_5 = 0.0 \quad a_5 = 1.0$$
$$b_6 = 0.0 \quad a_6 = 1.6$$
$$b_7 = 0.1 \quad a_7 = 1.6$$
$$b_8 = 0.2 \quad a_8 = 1.4$$
$$b_9 = 0.3 \quad a_9 = 1.1$$
$$b_{10} = 0.4 \quad a_{10} = 1.7$$

The R command lines are as follows:

```
> b <- c(-0.4, -0.3, -0.2, -0.1, 0.0, 0.0, 0.1, 0.2, 0.3, 0.4)
> a <- c(1.0, 1.5, 1.2, 1.3, 1.0, 1.6, 1.6, 1.4, 1.1, 1.7)
> theta <- seq(-3, 3, 0.1)
> J <- length(b)
> ii <- matrix(rep(0, length(theta)*J), nrow=length(theta))
> i <- rep(0, length(theta))
> for (j in 1:J) {
    P <- 1 / (1 + exp(-a[j] * (theta - b[j])))
    ii[,j] <- a[j]**2 * P * (1.0 - P)
    i <- i + ii[,j]
  }
> plot(theta, i, xlim=c(-3,3), ylim=c(0,10), type="l",
    xlab="Ability", ylab="Information",
    main="Test Information Function")
```

By executing the R command lines, the test information function in Fig. 6.5 will appear on the screen in the graphics window.

The first two command lines are used to set up the known item parameters for the ten items. The third line is used to set up the 61 ability points on the ability scale.

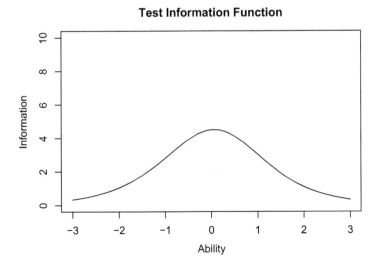

Fig. 6.5 Test information function

The number of items *J* will be determined in the fourth line based on the size of the vector contains the item difficulty parameters. The next two lines are used to set up the matrix that will contain the values of item information functions (i.e., `ii`) and the vector that will contain the values of test information function (i.e., `i`). The `for` loop in the next line performs the computation of item information from the first to the last item and sums up the item information to obtain the test information. The item information function for item *j* is contained in the column *j* of the matrix `ii` (i.e., `ii[,j]`). The test information function will be plotted by executing the last command line. Note that the information function will have only positive values. The maximum value of the vertical axis is arbitrarily set to 10 in the above example.

The test information function is symmetric about an ability level of 0.0, reflecting the distribution of the item difficulties around zero. The maximum value of the amount of test information is approximately 4.2, which yields a standard error of estimate of 0.49. Within the range of ability from -1.0 to $+1.0$ the amount of test information is greater than 2.5 and the standard error of estimate is less than 0.63 in this range. Outside of this range the amount of information is smaller, and at an ability level of -2.0 or $+2.0$ is only about 1.0. At these points, the standard error of estimate is 1.0. Since this test has only 10 items, the general level of the test information function is at a modest value and the precision reflects this.

In addition, using similar R command lines, a demonstration of the calculations of item information functions and the resulting test information function can be easily done. For example, values reported in Table 6.4 can be obtained from the following R command lines:

```
> b <- c(-1.0, -0.5, 0.0, 0.5, 1.0)
> a <- c(2.0, 1.5, 1.5, 1.5, 2.0)
> theta <- seq(-3, 3, 1)
> J <- length(b)
> ii <- matrix(rep(0, length(theta)*J), nrow=length(theta))
> i <- rep(0, length(theta))
> for (j in 1:J) {
    P <- 1 / (1 + exp(-a[j] * (theta - b[j])))
    ii[,j] <- a[j]**2 * P * (1.0 - P)
    i <- i + ii[,j]
  }
> theta; ii; i
```

6.7.2 An R Function for Test Information Functions

It is possible to create an R function for plotting a test information function given a set of item parameters. Consider the following function named tif:

```
> tif <- function(b, a, c) {
    J <- length(b)
    if (missing(c)) c <- rep(0, J)
```

6.8 Exercises

```
        if (missing(a)) a <- rep(1, J)
        theta <- seq(-3, 3, 0.1)
        ii <- matrix(rep(0, length(theta)*J), nrow=length(theta))
        i <- rep(0, length(theta))
        for (j in 1:J) {
           Pstar <- 1 / (1 + exp(-a[j] * (theta - b[j])))
           P <- c[j] + (1 - c[j]) * Pstar
           ii[,j] <- a[j]**2 * P * (1.0 - P) * (Pstar / P)**2
           i <- i + ii[,j]
        }
        plot(theta, i, xlim=c(-3,3), ylim=c(0,10), type="l",
           xlab="Ability", ylab="Information",
           main="Test Information Function")
     }
```

Note that use of the function tif requires the vectors of item parameters. Also note that the maximum of the information value on the vertical axis of the graph is arbitrarily set to 10.0.

For the three-parameter model, the actual equation used in the function tif is not Eq. (6.5) but an equivalent formula,

$$I_j(\theta) = a_j^2 P_j(\theta) Q_j(\theta) \left[\frac{P_j^*(\theta)}{P_j(\theta)}\right]^2,$$

where

$P_j^*(\theta) = 1/[1 + \exp(a_j(\theta - b_j))],$
$P_j(\theta) = c_j + (1 - c_j)P_j^*(\theta),$ and
$Q_j(\theta) = 1 - P_j(\theta).$

6.8 Exercises

For the following exercises, it is assumed that you have defined the function tif by typing it in the R console window.

6.8.1 Using the Two-Parameter Model

1. Obtain a plot of the test information function from a test of 10 items with the following specifications:

   (a) Set all the values of $b = 0.0$.
   (b) Use various values of a that are all greater than 1.0 but less than 1.7.

   The test information function will look quite similar to the one in the computer session.

2. Obtain a plot of the test information function from a test of 10 items with the following specifications:

    (a) Set all the values of $b = 0.0$.
    (b) Use various values of a that are all less than 1.0 but greater than 0.0.

    The test information function will symmetric about zero but will have a much lower overall level than the previous test information function. Note that you may use the R command line, par(new=T), to plot the new information function onto the existing graph.

3. Obtain a plot of the test information function from a test of 10 items with the following specifications:

    (a) Set all the values of $b = 0.0$.
    (b) Use various values of a that are all greater than 1.7. The maximum value you may use is 2.0.

    The test information function will have a maximum greater than that of all of the previous examples, thus illustrating the dependence of the amount of information upon the values of the item discrimination parameter.

4. Repeat one of the above examples using a test of five items. For example,

    (a) Set all the values of $b = 0.0$.
    (b) Use various values of a that are all greater than 1.7. The maximum value you may use is 2.0.

    The general level of the test information function will be much lower than the corresponding example. depending on how you choose the values of b and a, the shape of the curve could be quite similar to the previous case.

6.8.2 Using the Rasch Model

1. Obtain a plot of the test information function from a test of ten items with the following specification:

    (a) Set all the values of the difficulty parameter to some common value other than zero.

    The test information curve will be centered on this common value. The general level of the amount of information will be modest because the Rasch model fixed the discrimination parameter at 1.0.

2. Obtain a plot of the test information function from a test of ten items with the following specification:

    (a) Set all the values of b that are equally spaced over the full range of ability from -3 to $+3$.

    The test information function will be rather flat and the general amount of information will be rather low.

6.8.3 Using the Three-Parameter Model

1. Obtain a plot of the test information function from a test of 10 items with the following specifications:
   (a) Select values for b and a that vary in value.
   (b) Set the value of $c = 0.1$ for all items.
   (c) Write down the values of b and a so they can be used again.

   Take note of the shape and general level of the obtained test information function.
2. Obtain a plot of the test information function from a test of 10 items with the following specifications:
   (a) Use the same values of b and a as the previous problem.
   (b) Set all the values of $c = 0.35$.

   The resulting information function will have a shape similar to that of the previous problem. However, the general level of the amount of test information will be less than that of the previous example. This illustrates the effect of guessing upon the precision with which ability is estimated.

6.8.4 Further Exercises

1. Use a model of your choice and select values of the item parameters such that the test information function approximates a horizontal line. Use a ten-item test.
2. Experiment a bit with different item characteristic curve models, parameter values, and number of items. To make things easier, you may use R command lines that lets the computer generate the values of the item parameters.
3. Plot several test information functions on the same graph. It will be helpful to make rough sketches of the test information functions displayed and notes to indicate the nature of the mix of item parameter values. The goal is to be able to predict what the form of the test information function will be from the values of the item parameters.

6.9 Things to Notice

1. The general level of the test information function depends upon:
   (a) The number of items in the test.
   (b) The average value of the discrimination parameters of the test items.
   (c) Both of the above hold for all three item characteristic curve models.

2. The shape of the test information function depends upon:
   (a) The distribution of the item difficulties over the ability scale.
   (b) The distribution and the average value of the discrimination parameters of the test items.
3. When the item difficulties are clustered closely around a given value, the test information function is peaked at that point on the ability scale. The maximum amount of information depends upon the values of the discrimination parameters.
4. When the item difficulties are widely distributed over the ability scale, the test information function tends to be flatter than when the difficulties are tightly clustered.
5. Values of $a < 1.0$ result in a low general level of the amount of test information.
6. Values of $a > 1.7$ result in a high general level of the amount of test information.
7. Under the three-parameter model, values of the guessing parameter c greater than zero lower the amount of test information at the low ability levels. In addition, large values of c reduce the general level of the amount of test information.
8. It is difficult to approximate a horizontal test information function. To do so, the values of b must be spread widely over the ability scale and the values of a must be in the moderate to low range and have a U-shaped distribution.

Chapter 7
Test Calibration

7.1 Introduction

For didactic purposes, all of the preceding chapters have assumed that the metric of the ability scale was known. This metric had a midpoint of zero, a unit of measurement of 1, and a range from negative infinity to positive infinity. The numerical values of the item parameters and the examinee's ability parameters have been expressed in this metric. While this has served to introduce you to the fundamental concepts of item response theory, it does not represent the actual testing situation. When test constructors write an item, they know what trait they want the item to measure and whether the item is designed to function among low-, medium-, or high-ability examinees. But it is not possible to determine the values of the item's parameters a priori. In addition, when a test is administered to a group of examinees, it is not known in advance how much of the latent trait each of the examinees possesses. As a result, a major task is to determine the values of the item parameters and examinee abilities in a metric for the underlying latent trait. In item response theory, this task is called test calibration and it provides a frame of reference for interpreting test results. Test calibration is accomplished by administering a test to a group of N examinees and dichotomously scoring the examinees' responses to the J items. Then mathematical procedures are applied to the item response data in order to create an ability scale that is unique to the particular combination of test items and examinees. The values of the item parameter estimates and the examinees' estimated abilities are expressed in this metric. Once this is accomplished, the test has been calibrated and the test results can be interpreted via the constructs of item response theory.

7.2 The Test Calibration Process

In 1968, Allan Birnbaum proposed a paradigm for calibrating a test under item response theory. This paradigm has been implemented in two widely used computer programs BICAL (Wright and Mead 1976) and LOGIST (Wingersky et al. 1982). The direct descendant of BICAL is the computer program WINSTEPS (Linacre 2015). Also the personal computer version of LOGIST, PCLOGIST (Wingersky et al. 1999) is available. The Birnbaum paradigm is an iterative procedure employing two stages of maximum likelihood estimation and is typically referred to as the joint maximum likelihood estimation procedure (Baker 1992; Baker and Kim 2004). In one stage, the parameters of the J items in the test are estimated, and in the second stage, the ability parameters of the N examinees are estimated. The two stages are performed iteratively until a stable set of parameter estimates is obtained. At this point, the test has been calibrated and an ability scale metric defined.

Within the first stage of the Birnbaum paradigm, the estimated ability of each examinee is treated as if it is expressed in the true metric of the latent trait. Then the parameters of each item in the test are estimated via the maximum likelihood procedure discussed in Chap. 3. This is done one item at a time, as an underlying assumption is that the items are independent of each other. The result is a set of values for the estimates of the parameters of the items in the test.

The second stage assumes that the item parameter estimates yielded by the first stage are actually the values of the item parameters. Then, the ability of each examinee is estimated using the maximum likelihood procedure presented in Chap. 5. It is assumed that the ability of each examinee is independent of all other examinees. Hence, the ability estimates are obtained one examinee at a time.

The two-stage process is repeated until some suitable convergence criterion is met. The overall effect is that the parameters of the J test items and the ability levels of the N examinees have been estimated simultaneously even though they were done one at a time. This clever paradigm reduces a very complex estimation problem to one that can be implemented on a computer.

7.2.1 The Metric Problem

An unfortunate feature of the Birnbaum paradigm is that it does not yield a unique metric for the ability scale. That is, the midpoint and the unit of measurement of the obtained ability scale are indeterminate; that is, many different values work equally well. In technical terms, the metric is unique up to a linear transformation. As a result, it is necessary to "anchor" the metric via arbitrary rules for determining the midpoint and the unit of measurement of the ability scale. How this is done is up to the persons implementing the Birnbaum paradigm in a computer program. In the BICAL and WINSTEPS computer programs, this anchoring process is performed after the first stage is completed. Thus, each of two stages within an iteration is performed using a slightly different ability scale metric. As the overall iterative process converges, the metric of the ability scale also converges to a particular midpoint and a unit of measurement. The crucial feature of this process is that the resulting ability scale metric depends upon the specific set of items constituting the test and the responses of a particular group of examinees to that test. It is not possible to obtain estimates of the examinee's ability and of the item's parameters in the true metric of the underlying latent trait. The best we can do is obtain a metric that depends upon a particular combination of examinees and test items.

7.3 Test Calibration Under the Rasch Model

There are three different item characteristic curve models to choose from and several different ways to implement the Birnbaum paradigm. From these, the authors has chosen to present the approach based upon the Rasch logistic item characteristic curve model as implemented by Benjamin D. Wright and his co-workers in the BICAL computer program. Under this model, only the item difficulty parameter is estimated for each item. The estimation procedures work well with small numbers of test items and small numbers of examinees. The metric anchoring procedure is simple and the basic ideas of test calibration are easy to present.

The calibration of a ten-item test administered to a group of 16 examinees will be used below to illustrate the process. The information presented is based upon the analysis of data set 1 contained in the computer session. You may elect to work through this section in parallel with the computer session, but it is not necessary as all the computer displays will be presented in the text.

The ten-item test is one that has been matched to the average ability of a group of 16 examinees. The examinees' item responses have been dichotomously scored, 1 for correct and 0 for incorrect. The goal is to use this item response data to calibrate the test. The actual item response vectors for each examinee are presented in Table 7.1 and each row represents the item responses made by a given examinee.

In Chap. 5 it was observed that it is impossible to estimate an examinee's ability by the maximum likelihood procedure if he or she gets none or all of the test items correct. Inspection of Table 7.1 reveals that examinee 16 answered all of the items correctly and must be removed from the data set. Similarly, if an item is answered correctly by all of the examinees or by none of the examinees, its item difficulty parameter cannot be estimated. Hence, such an item must be removed from the data set. In this particular example, no items were removed for this reason. One of the unique features of test calibration under the Rasch model is that all examinees having the same number of items correct (the same raw score) will obtain the same estimated ability. As a result, it is not necessary to distinguish among the several examinees having the same raw test score. Consequently, rather than using the individual item responses, all that is needed is the number of examinees at each raw score answering each item correctly. Because of this and the removing of items, an edited data set is used as the initial starting point for test calibration procedures under the Rasch model. The edited data set for this example is presented in Table 7.2.

Table 7.1 Item responses by examinees for the matched test

Examinee	Item 1	2	3	4	5	6	7	8	9	10	Raw score
1	0	0	1	0	0	0	0	1	0	0	2
2	1	0	1	0	0	0	0	0	0	0	2
3	1	1	1	0	1	0	1	0	0	0	5
4	1	1	1	0	1	0	0	0	0	0	4
5	0	0	0	0	1	0	0	0	0	0	1
6	1	1	0	1	0	0	0	0	0	0	3
7	1	0	0	0	0	1	1	1	0	0	4
8	1	0	0	0	1	1	0	0	1	0	4
9	1	0	1	0	0	1	0	0	1	0	4
10	1	0	0	0	1	0	0	0	1	0	3
11	1	1	0	1	1	1	1	1	1	1	9
12	1	1	1	1	1	1	1	1	1	0	9
13	1	1	1	0	1	0	1	0	0	1	6
14	1	1	1	1	1	1	1	1	1	0	9
15	1	1	0	1	1	1	1	1	1	1	9
16	1	1	1	1	1	1	1	1	1	1	10

Table 7.2 Frequency counts for the edited data (examinee no. 16 eliminated) for the matched test

Score	Item 1	2	3	4	5	6	7	8	9	10	Row sum	Score frequency
1	0	0	0	0	1	0	0	0	0	0	1	1
2	1	0	2	0	0	0	0	1	0	0	4	2
3	2	1	0	1	1	0	0	0	1	0	6	2
4	4	1	2	0	2	3	1	1	2	0	16	4
5	1	1	1	0	1	0	1	0	0	0	5	1
6	1	1	1	0	1	0	1	0	0	1	6	1
7	0	0	0	0	0	0	0	0	0	0	0	0
8	0	0	0	0	0	0	0	0	0	0	0	0
9	4	4	2	4	4	4	4	4	4	2	36	4
Col. sum	13	8	8	5	10	7	7	6	7	3	74	15

In Table 7.2, the rows are labeled by raw test scores ranging from 1 to 9. The row marginals (i.e., row sums) are the total number of correct responses made by examinees with that raw test score. By dividing each row sum by the corresponding score we can obtain the frequency of the examinees for each score. The columns are labeled by the item number from 1 to 10. The column marginals (i.e., column sums) are the total number of correct responses made to the particular item by the remaining examinees. Under the Rasch model, the only information used in the Birnbaum paradigm are the frequency totals contained in the row and column

7.3 Test Calibration Under the Rasch Model

marginals. This is unique to this model and results in simple computations within the maximum likelihood estimation procedures employed at each stage of the overall process.

Given the two frequency vectors (i.e., the column sum vector and the score frequency vector), the estimation process can be implemented. Initial estimates are obtained for the item difficulty parameters for the first stage and the metric of the ability scale must be anchored. Under the Rasch model the anchoring procedure takes advantage of the fact that the item discrimination parameter is fixed at a value of one for all items in the test. Because of this, the unit of measurement of the estimated abilities is fixed at a value of 1. All that remains then is to define the midpoint of the scale. In the BICAL computer program, the midpoint is defined as the mean of the estimated item difficulties. In order to have a convenient midpoint value, the mean item difficulty is subtracted from the value of each item's difficulty estimate, resulting in the rescaled mean item difficulty having a value of zero. Because the item difficulties are expressed in the same metric as the ability scale, the midpoint and the unit of measurement of the latter have now been determined. Since this is done between stages, the abilities estimated in the second stage will be in the metric defined by the rescaled item parameter estimates obtained in the first stage.

The ability estimate corresponding to each raw test score is obtained in the second stage using the rescaled item difficulty estimates as if they were the item difficulty parameters and the vector of the score frequencies. The output of this stage is an ability estimate for each raw test score in the data set. At this point, the convergence of the overall iterative process is checked. In the BICAL program, Wright summed the absolute differences between the values of the item difficulty parameter estimates for two successive iterations of the paradigm. If this sum was less than 0.01, the estimation process was terminated. If it was greater than 0.01, then another iteration was performed and the two stages were done again. Thus, the process of (1) stage one, (2) anchoring the metric, (3) stage two, and (4) convergence check is repeated until the criterion is met. When this happens, the current values of the item and ability parameter estimates are accepted and an ability scale metric has been defined. The estimates of the item difficulty parameters for the present example are presented in Table 7.3.

You can verify that the sum of the item difficulties is zero (within rounding errors). The interpretation of the values of the item parameter estimates is exactly that presented in Chap. 2. For example, item 1 has an item difficulty of −2.37, which locates it at the low end of the ability scale. Item 6 has a difficulty of 0.11, which locates it near the middle of the ability scale. Item 10 has a difficulty of 2.06, which locates it at the high end of the ability scale. Thus, the usual interpretation of item difficulty as locating the item on the ability scale holds. Because of the anchoring procedures used, these values are actually relative to the average item difficulty of the test for these examinees.

The ability estimate has been reported in Table 7.4 for each score. All examinees with the same raw score obtained the same ability estimate. For example, examinees 1 and 2 both had raw scores of 2 and obtained an estimated ability of −1.50.

Table 7.3 Estimated item difficulty parameters for the matched test

Item	Difficulty
1	−2.37
2	−0.27
3	−0.27
4	0.98
5	−1.00
6	0.11
7	0.11
8	0.52
9	0.11
10	2.06

Table 7.4 Obtained ability estimates for the matched test

Raw score	Ability	Score frequency	Examinee no.
1	−2.37	1	5
2	−1.50	2	1, 2
3	−0.91	2	6, 10
4	−0.42	4	4, 7, 8, 9
5	0.02	1	3
6	0.46	1	13
7	0.93	0	
8	1.50	0	
9	2.32	4	11, 12, 14, 15

Note: Examinee 16 has been eliminated

Examinees 4, 7, 8, and 9 had raw scores of 4 and shared a common estimated ability of −0.42. This unique feature is a direct consequence of the fact that, under the Rasch model the value of the discrimination parameter is fixed at 1 for all of the items in the test. This aspect of the Rasch model is appealing to practitioners as they intuitively feel that examinees obtaining the same raw test score should receive the same ability estimate. When the two- and three-parameter item characteristic curve models are used, an examinee's ability estimate depends upon the particular pattern of item responses rather than the raw score. Under these models, examinees with the same item response pattern will obtain the same ability estimate. Thus, examinees with the same raw score could obtain different ability estimates if they answered different items correctly.

Examinee 16 was not included in the computations due to being removed because of a perfect raw score. The ability estimate obtained by a given examinee is interpreted in terms of where it locates the examinee on the ability scale. For example, examinee 7 had an estimated ability of −0.42 which places him or her just below the midpoint of the scale. The ability estimates can be treated just like any other score. Their distribution over the ability scale can be plotted and the summary statistics of this distribution can be computed. By weighting the ability estimates with the respective score frequencies, in the present case this yields a mean of 0.06

and a standard deviation of 1.57. Thus, examinee 7 had an ability score that was about 0.23 standard deviations below the mean ability of the group. However, one would not typically interpret an examinee's ability score in terms of the distribution of the scores for the group of examinees. To do so is to ignore the fact that the ability score can be interpreted directly as the examinee's position on the ability scale.

7.4 Summary of the Test Calibration Process

The end product of the test calibration process is the definition of an ability scale metric. Under the Rasch model, this scale has a unit of measurement of 1 and a midpoint of zero. Superficially this looks exactly the same as the ability scale metric used in previous chapters. However, it is not the metric of the underlying latent trait. The obtained metric depends upon the item responses yielded by a particular combination of examinees and test items being subjected to the Birnbaum paradigm. Since the true metric of the underlying latent trait cannot be determined, the metric yielded by the Birnbaum paradigm is used as if it were the true metric. The obtained item difficulty values and the examinee's ability are interpreted in this metric. Thus, the test has been calibrated. The outcome of the test calibration procedure is to locate each examinee and item along the obtained ability scale. In the present example, item 5 had a difficulty of -1.00 and examinee 10 had an ability estimate of -0.91. Therefore, the probability of examinee 10 answering item 5 correctly is slightly less than 0.5. The capability to locate items and examinees along a common scale is a powerful feature of item response theory. This feature allows one to interpret the results of a test calibration within a single framework and provides meaning to the values of the parameter estimates.

7.5 Computer Session

This computer session is a bit different from those of the previous chapters. Because it would be difficult for you to create data sets to be calibrated, three sets have been created for an illustration purpose. Each of these will be used to calibrate a test and the results will be displayed on the screen. You will simply step through each of the data sets and calibration results. There are some definite goals in this process. First, you will become familiar with the input data and how it is edited. Second, the item difficulty estimates and the examinee's ability estimates can be interpreted. Third, the test characteristic curve and test information functions for the test will be shown and interpreted.

Three different ten-item tests measuring the same latent trait will be used. A common group of 16 examinees will take all three of the tests. The tests were created so that the average difficulty of the first test was matched to the mean ability of the common group of examinees. The second test was created to be an easy test for this group. The third test was created to be a hard test for this group. Each of these test group combinations will be subjected to the Birnbaum paradigm and calibrated separately. There are two reasons for this approach. First, it illustrates that each test calibration yields a unique metric for the ability scale. Second, the results can be used to show the process by which the three sets of test results can be placed on a common ability scale.

7.5.1 Procedures for the Test Calibration Session: Data Set 1

The following R command lines are needed to calibrate data set 1 for the matched test presented in Table 7.1:

```
> rm(list = ls())
> s <- c(13, 8, 8, 5, 10, 7, 7, 6, 7, 3)
> f <- c(1, 2, 2, 4, 1, 1, 0, 0, 4)
> convb <- 0.01; convt <- 0.01; convabd <- 0.01
> J <- length(s); G <- length(f); K <- 25; T <- 10
> b <- log((sum(f) - s) / s)
> b <- b - mean(b)
> oldb <- b
> theta <-seq(1, G, 1)
> for (g in 1:G) { theta[g] <- log(g / (J - g)) }
> for (k in 1:K) {
    cat("cycle k=", k, "\n")
    for (j in 1:J) {
      for (t in 1:T) {
        sumfp <- 0
        sumfpq <- 0
        for (g in 1:G) {
          p <- 1 / (1 + exp(-(theta[g] - b[j])))
          sumfp <- sumfp + f[g] * p
          sumfpq <- sumfpq + f[g] * p * (1 - p)
        }
        deltab <- (s[j] - sumfp) / sumfpq
        b[j] <- b[j] - deltab
        if (abs(deltab) < convb) { break }
      }
    }
    b <- b - mean(b)
    for (g in 1:G) {
      for (t in 1:T) {
        sump <- 0
        sumpq <- 0
```

7.5 Computer Session

```
            for (j in 1:J) {
               p <- 1 / (1 + exp(-(theta[g] - b[j])))
               sump <- sump + p
               sumpq <- sumpq - p * (1 - p)
            }
            deltat <- (g - sump) / sumpq
            theta[g] <- theta[g] - deltat
            if (abs(deltat) < convt) { break }
         }
      }
      abd <- abs(b - oldb)
      if (sum(abd) < convabd) { break }
      else { oldb <- b }
   }
> b <- b * ((J - 1) / J)
> for (j in 1:J) {
     cat("b(", j, ")=", b[j], "\n")
  }
> for (g in 1:G) {
     for (t in 1:T) {
        sump <- 0
        sumpq <- 0
        for (j in 1:J) {
           p <- 1 / (1 + exp(-(theta[g] - b[j])))
           sump <- sump + p
           sumpq <- sumpq - p * (1 - p)
        }
        deltat <- (g - sump) / sumpq
        theta[g] <- theta[g] - deltat
        if (abs(deltat) < convt) { break }
     }
  }
> theta <- theta * ((J - 2) / (J - 1))
> for (g in 1:G) {
     cat("theta(", g, ")=", theta[g], "\n")
  }
```

The first line removes the existing objects in workspace. Note that after obtaining item and ability parameter estimates from the Birnbaum paradigm, the bias correction methods were applied to the item parameter estimates and then to the ability estimates.

This ten-item test has a mean difficulty that is matched to the average ability of the group of 16 examinees. Item response vectors of the 16 examinees are presented in Table 7.1. Notice that examinee 16 answered all items correctly. After deleting examinee 16, the edited data are presented in Table 7.2. Notice that examinee 16 has been eliminated and that no items were eliminated for data set 1. The vector that contains the column sum for the ten items, s, and the vector that contains the frequencies for the scores from 1 to 9, f, are the input for the Birnbaum paradigm to calibrate the test.

By executing the R command lines, the R console window shows the item difficulty estimates for the matched test, for example:

```
b( 1 )= -2.36761
b( 2 )= -0.265167
b( 3 )= -0.265167
b( 4 )= 0.9763713
b( 5 )= -0.9975242
b( 6 )= 0.1127705
b( 7 )= 0.1127705
b( 8 )= 0.5210009
b( 9 )= 0.1127705
b( 10 )= 2.059785
```

The values are the same as those in Table 7.3. The estimated abilities are not for individual examinees but for the raw score groups ranged from 1 to 9:

```
theta( 1 )= -2.370017
theta( 2 )= -1.499325
theta( 3 )= -0.9058965
theta( 4 )= -0.4206984
theta( 5 )= 0.02104907
theta( 6 )= 0.4593398
theta( 7 )= 0.9328307
theta( 8 )= 1.501894
theta( 9 )= 2.328257
```

The ability estimates are the same as those reported in Table 7.4. The ability estimates of the 15 examinees had a mean of 0.06 and a standard deviation of 1.57. Notice that examinee 16 did not receive an ability estimate.

Using the item difficulty parameter estimates, the test characteristic curve can be calculated and displayed in the graphics window for data set 1 (see data set 1 of Fig. 7.1). Take note of the fact that the mid true score (a true score equal to one-half the number of items) corresponds to an ability level of 0. This reflects the anchoring procedure that sets the average item difficulty to zero.

The test information function can also be constructed and displayed in the graphics window for data set 1 (see data set 1 of Fig. 7.2). The curve is reasonably symmetric and has a well-defined hump in the middle. The form of the curve indicates that ability is estimated with the greatest precision in the neighborhood of the middle of the ability scale. The peak of the test information function occurs at

7.5 Computer Session

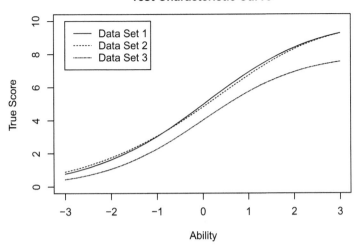

Fig. 7.1 Test characteristic curves for the three data sets

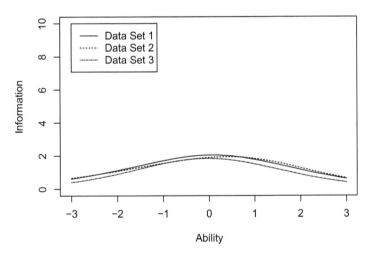

Fig. 7.2 Test information functions for the three data sets

a point slightly above the midpoint of the ability scale. This reflects the distribution of the item difficulty estimates, as there were six items with positive values and only four with negative values. Thus, there is a very slight emphasis upon positive ability levels. The maximum amount of information is roughly 2, which is rather small.

Instead of using the column sums and the score frequencies from the edited data, the following R command lines can be used to obtain s and f from the original data by eliminating the examinees with either 0 or perfect score:

```
> N <- 16
> J <- 10
> U <- matrix(c(
    0, 0, 1, 0, 0, 0, 0, 1, 0, 0,
    1, 0, 1, 0, 0, 0, 0, 0, 0, 0,
    1, 1, 1, 0, 1, 0, 1, 0, 0, 0,
    1, 1, 1, 0, 1, 0, 0, 0, 0, 0,
    0, 0, 0, 0, 1, 0, 0, 0, 0, 0,
    1, 1, 0, 1, 0, 0, 0, 0, 0, 0,
    1, 0, 0, 0, 0, 1, 1, 1, 0, 0,
    1, 0, 0, 0, 1, 1, 0, 0, 1, 0,
    1, 0, 1, 0, 0, 1, 0, 0, 1, 0,
    1, 0, 0, 0, 1, 0, 0, 0, 1, 0,
    1, 1, 0, 1, 1, 1, 1, 1, 1, 1,
    1, 1, 1, 1, 1, 1, 1, 1, 1, 0,
    1, 1, 1, 0, 1, 0, 1, 0, 0, 1,
    1, 1, 1, 1, 1, 1, 1, 1, 1, 0,
    1, 1, 0, 1, 1, 1, 1, 1, 1, 1,
    1, 1, 1, 1, 1, 1, 1, 1, 1, 1
    ), nrow=N, ncol=J, byrow=TRUE)
> f <- unname(table(factor(sort(rowSums(U)),
    levels=0:J)))
> f <- f[-1]
> nperfect <- f[J]
> f <- f[-J]
> s <- colSums(U) - nperfect
```

7.5.2 Data Set 2

This ten-item test was constructed to be an easy test for the common group of 16 examinees. Since the computer procedures for this data set will be exactly the same as for data set 1, they will not be repeated in detail. Only the significant results will be noted. For data set 2, item responses of the ten items by 16 examinees are presented in Table 7.5.

7.5 Computer Session

Table 7.5 Item responses by examinees for the easy test

Examinee	Item 1	2	3	4	5	6	7	8	9	10	Raw score
1	0	0	1	0	0	0	0	0	0	0	1
2	0	1	1	1	0	0	0	1	0	0	4
3	1	0	0	1	0	0	0	0	0	0	2
4	1	1	1	0	0	0	1	0	0	1	5
5	1	1	1	1	0	0	0	0	0	1	5
6	1	1	1	0	1	1	0	0	0	1	6
7	1	1	1	1	1	1	1	1	1	0	9
8	1	1	1	0	1	1	1	0	0	0	6
9	0	1	1	1	1	1	0	1	1	0	7
10	1	0	1	1	1	1	1	0	1	0	7
11	1	1	1	1	1	0	0	1	0	1	7
12	1	1	1	1	1	0	1	1	1	0	8
13	1	1	1	0	1	1	0	0	0	0	5
14	1	1	1	0	1	1	1	1	1	1	9
15	1	1	1	1	1	1	1	1	1	1	10
16	1	1	1	1	1	1	1	1	1	1	10

The R command lines to obtain the vector of column sums and the vector of frequencies of scores from 1 to 9 by removing both zero and perfect scores are listed below:

```
> N <- 16
> J <- 10
> U <- matrix(c(
    0, 0, 1, 0, 0, 0, 0, 0, 0, 0,
    0, 1, 1, 1, 0, 0, 0, 1, 0, 0,
    1, 0, 0, 1, 0, 0, 0, 0, 0, 0,
    1, 1, 1, 0, 0, 0, 1, 0, 0, 1,
    1, 1, 1, 1, 0, 0, 0, 0, 0, 1,
    1, 1, 1, 0, 1, 1, 0, 0, 0, 1,
    1, 1, 1, 1, 1, 1, 1, 1, 1, 0,
    1, 1, 1, 0, 1, 1, 1, 0, 0, 0,
    0, 1, 1, 1, 1, 1, 0, 1, 1, 0,
    1, 0, 1, 1, 1, 1, 1, 0, 1, 0,
    1, 1, 1, 1, 1, 0, 0, 1, 0, 1,
    1, 1, 1, 1, 1, 0, 1, 1, 1, 0,
    1, 1, 1, 0, 1, 1, 0, 0, 0, 0,
    1, 1, 1, 0, 1, 1, 1, 1, 1, 1,
    1, 1, 1, 1, 1, 1, 1, 1, 1, 1,
    1, 1, 1, 1, 1, 1, 1, 1, 1, 1
), nrow=N, ncol=J, byrow=TRUE)
```

```
> f <- unname(table(factor(sort(rowSums(U)), levels=0:J)))
> f <- f[-1]
> nperfect <- f[J]
> f <- f[-J]
> s <- colSums(U) - nperfect
```

Executing the command lines and by typing in the variable names, we can obtain the values of s and f as follows:

```
> f
[1]  1  1  0  1  3  2  3  1  2
> s
[1] 11 11 13  8  9  7  6  6  5  5
```

The values of s and f are based on the edited data; examinees 15 and 16 have been eliminated for having perfect raw scores.

The obtained item parameter estimates and the ability estimates for the score groups from 1 to 9 are as follows:

```
b(  1 ) = -1.112752
b(  2 ) = -1.112752
b(  3 ) = -2.752128
b(  4 ) =  0.2047593
b(  5 ) = -0.1761493
b(  6 ) =  0.5630637
b(  7 ) =  0.9155878
b(  8 ) =  0.9155878
b(  9 ) =  1.277391
b( 10 ) =  1.277391

theta( 1 ) = -2.525016
theta( 2 ) = -1.579138
theta( 3 ) = -0.9235425
theta( 4 ) = -0.388803
theta( 5 ) =  0.08882409
theta( 6 ) =  0.5474272
theta( 7 ) =  1.023334
theta( 8 ) =  1.57362
theta( 9 ) =  2.355977
```

The mean of the estimated item difficulties is zero. Six of the items obtained positive item difficulty estimates and the distribution of the difficulties is somewhat U-shaped. By using the score frequencies and the respective ability estimates, the ability estimates of the 14 examinees had a mean of 0.44 and a standard deviation of 1.35. It is interesting to note that examinee 9 had a raw score of 4 on the matched test and obtained an estimated ability of -0.42. On this easy test, the raw score was 7 and the ability estimate was 1.02. Yet the examinee's true ability is the same in both cases.

7.5 Computer Session

Table 7.6 Item responses by examinees for the hard test

Examinee	Item										Raw score
	1	2	3	4	5	6	7	8	9	10	
1	0	0	0	0	0	0	0	0	0	0	0
2	1	0	0	0	0	0	0	0	0	0	1
3	0	0	0	0	0	0	0	0	0	0	0
4	1	0	0	0	0	1	1	0	0	0	3
5	1	0	0	1	0	0	1	0	0	0	3
6	1	0	0	0	0	0	0	0	0	0	1
7	1	0	1	0	0	0	0	0	0	0	2
8	1	0	1	1	0	0	0	0	0	0	3
9	1	0	0	0	0	0	1	0	0	0	2
10	1	0	0	1	0	0	0	1	0	0	3
11	1	1	1	0	0	1	0	0	1	0	5
12	1	1	1	1	1	1	1	1	0	0	8
13	1	1	0	0	1	1	0	0	1	0	5
14	1	1	1	1	1	1	1	1	0	0	8
15	1	1	1	0	1	1	0	0	1	0	7
16	1	1	1	0	0	1	1	1	1	0	8

The mid true score of the test characteristic curve again corresponds to an ability level of zero. The form of the test characteristic curve is nearly identical to that of the first test (see Fig. 7.1). The test information function is symmetric and has a somewhat rounded appearance (see Fig. 7.2). The maximum amount of information (2.0) occurred at an ability level of roughly 0.5.

7.5.3 Data Set 3

This ten-item test was constructed to be a hard test for the common group of 16 examinees. Because the computer procedures will be similar to the previous two examples, only the results of interest will be discussed. Table 7.6 contains the item responses to the hard test of ten items by the 16 examinees.

Inspection of the table of item response vectors shows that examinees 1 and 3 have raw scores of zero and will be removed. Inspection of the columns reveals that none of the examinees answered item 10 correctly and it will be removed from the data set. In addition, after removing the two examinees, item 1 was answered correctly by all of the remaining examinees. Thus, this item must also be removed. Upon doing this, examinees 2 and 6 now have raw scores of zero as the only item they answered correctly was item 1. After removing these two additional examinees no further editing is needed. Such multiple-stage editing is quite common in test calibrating. It should be noted that after editing the data set is smaller than the previous two and the range of raw scores is now from 1 to 7.

Due to the iterative multiple-stage editing required for data set 3, somewhat different R command lines are needed to obtain the final vector of the column sums and the vector of the score frequencies for scores 1 to 7. The R command lines for data set 3 with a bit more general data editing lines to get the two vectors of s and f are as follows:

```
> N <- 16
> J <- 10
> U <- matrix(c(
    0, 0, 0, 0, 0, 0, 0, 0, 0, 0,
    1, 0, 0, 0, 0, 0, 0, 0, 0, 0,
    0, 0, 0, 0, 0, 0, 0, 0, 0, 0,
    1, 0, 0, 0, 0, 1, 1, 0, 0, 0,
    1, 0, 0, 1, 0, 0, 1, 0, 0, 0,
    1, 0, 0, 0, 0, 0, 0, 0, 0, 0,
    1, 0, 1, 0, 0, 0, 0, 0, 0, 0,
    1, 0, 1, 1, 0, 0, 0, 0, 0, 0,
    1, 0, 0, 0, 0, 1, 0, 0, 0, 0,
    1, 0, 0, 1, 0, 0, 0, 0, 1, 0,
    1, 1, 1, 0, 0, 1, 0, 0, 1, 0,
    1, 1, 1, 1, 1, 1, 1, 1, 0, 0,
    1, 1, 0, 0, 1, 1, 0, 0, 1, 0,
    1, 1, 1, 1, 1, 1, 1, 1, 0, 0,
    1, 1, 1, 0, 1, 1, 0, 0, 1, 0,
    1, 1, 1, 0, 0, 1, 1, 1, 1, 0
  ), nrow=N, ncol=J, byrow=TRUE)
> u <- U
> oldu <- u
> for (j in 1:J) {
    s <- colSums(u)
    for (j in length(u[1,]):1) {
      if (s[j] == 0 | s[j] == length(u[,1])) {
        u <- u[,-j]
      }
    }
    f <- rowSums(u)
    for (i in length(u[,1]):1) {
      if (f[i] == 0 | f[i] == length(u[1,])) {
        u <- u[-i,]
      }
```

7.5 Computer Session

```
        }
        if (length(oldu[,1]) == length(u[,1]) &
            length(oldu[1,]) == length(u[1,])) { break }
        else { oldu <- u }
    }
> f <- unname(table(factor(sort(rowSums(u)),
    levels=1:(length(u[1,])-1))))
> s <- colSums(u)
```

The above lines yielded two vectors of s and f as:

```
> s
[1] 6 7 5 4 7 6 3 5
> f
[1] 2 4 0 2 1 1 2
```

The two vectors are to be used as the input to the Birnbaum paradigm procedure. The R command lines for the calibration yielded the following estimates:

b(1) = -0.29298
b(2) = -0.7132503
b(3) = 0.1409872
b(4) = 0.6059282
b(5) = -0.7132503
b(6) = -0.29298
b(7) = 1.124558
b(8) = 0.1409872

theta(1) = -1.552674
theta(2) = -0.8901916
theta(3) = -0.4204405
theta(4) = -0.006298059
theta(5) = 0.4102034
theta(6) = 0.8870681
theta(7) = 1.561365

The mean of the eight estimated item difficulties was zero. Four of the items had positive values of item difficulty estimates. Note that item numbers are based on the edited data, not based on the original ten items. The original item 8 (item 7 from the edited data) had a difficulty of 1.12 while the remaining seven item difficulties fell in the range of -0.71 to $+0.61$. The 12 examinees used in the test calibration had a mean of -0.22 and a standard deviation of 1.26.

The test characteristic curve is similar to the previous two and the mid true score occurs again at an ability level of zero (see Fig. 7.1). But the upper part of the curve approaches a value of 8 rather than 10. The test information function was nearly symmetrical about an ability level of roughly 0. The curve was a bit less peaked than either of the two previous test information functions and its maximum of about 1.8 was slightly lower.

The reader should ponder a bit as to why the mean ability of the common group of examinees is not the same for all three calibrations. The item invariance principle says that they should all be the same. Is the principle wrong or is something else functioning here? The resolution of this inconsistency is presented after the Things to Notice section.

7.5.4 An R Function for Calibration of the Rasch Model

It is possible to create an R function for calibration under the Birnbaum paradigm for the Rasch model. Consider the following function named rasch which requires two other functions, stage1 and stage2:

```
>   rm(list = ls())

>   rasch <- function(s, f) {
      J <- length(s); G <- length(f); K <- 25; T <- 10
      b <- log((sum(f) - s) / s)
      b <- b - mean(b)
      oldb <- b
      theta <-seq(1, G, 1)
      for (g in 1:G) {theta[g] <- log(g / (J - g)) }
      for (k in 1:K) {
        convabd <- 0.01
        cat("cycle k=", k, "\n")
        b <-  stage1(b, theta, s, f)
        b <- b - mean(b)
        theta <-  stage2(theta, b)
        abd <- abs(b - oldb)
        if (sum(abd) < convabd) { break }
        else { oldb <- b }
      }
      b <- b * ((J - 1) / J)
      for (j in 1:J) {
        cat("b(", j, ")=", b[j], "\n")
      }
      cat("mean(b)=", mean(b), "\n")
      cat("sd(b)=", sd(b), "\n")
      cat("J=", J, "\n")
      theta <- stage2(theta,b)
      theta <- theta * ((J - 2) / (J - 1))
      for (g in 1:G) {
        cat("theta(", g, ")=", theta[g], "\n")
      }
      cat("mean(theta)=", mean(rep(theta, f)), "\n")
      cat("sd(theta)=", sd(rep(theta, f)), "\n")
```

7.5 Computer Session

```
      cat("N=", sum(f), "\n")
      cat("f=", f, "\n")
    }

> stage1 <- function(b, theta, s, f) {
    J <- length(b); G <- length(theta); T <- 10
    for (j in 1:J) {
       convb <- 0.01
       for (t in 1:T) {
          sumfp <- 0
          sumfpq <- 0
          for (g in 1:G) {
             p <- 1 / (1 + exp(-(theta[g] - b[j])))
             sumfp <- sumfp + f[g] * p
             sumfpq <- sumfpq + f[g] * p * (1 - p)
          }
          deltab <- (s[j] - sumfp) / sumfpq
          b[j] <- b[j] - deltab
          if (abs(deltab) < convb) { break }
       }
    }
    return(b)
  }

> stage2 <- function(theta, b) {
    G <- length(theta); J <- length(b); T <- 10
    for (g in 1:G) {
       convt <- 0.01
       for (t in 1:T) {
          sump <- 0
          sumpq <- 0
          for (j in 1:J) {
             p <- 1 / (1 + exp(-(theta[g] - b[j])))
             sump <- sump + p
             sumpq <- sumpq - p * (1 - p)
          }
          deltat <- (g - sump) / sumpq
          theta[g] <- theta[g] - deltat
          if (abs(deltat) < convt) { break }
       }
    }
    return(theta)
  }
```

Note that blank lines are used to separate the functions.

After defining the functions, you can obtain the item and ability estimates by setting up the input and executing the function rasch. For example, for data 1 the following command lines will perform the calibration of data under the Rasch model and print out the item and ability estimates in the R console window:

```
s <- c(13, 8, 8, 5, 10, 7, 7, 6, 7, 3)
f <- c(1, 2, 2, 4, 1, 1, 0, 0, 4)
rasch(s,f)
```

Note that this function can also be used with the data editing command lines. Note also that the function rasch will also print out the mean and the standard deviation of the item parameter estimates as well as those of the ability estimates.

7.6 Exercises

For the following exercises, it is assumed that you have defined the function rasch by typing it in the R console window.

1. Obtain item and ability estimates for data 1 under the Rasch model.
2. Obtain item and ability estimates for data 2 under the Rasch model.
3. Obtain item and ability estimates for data 3 under the Rasch model.

7.7 Things to Notice

1. In all three calibrations, examinees were removed in the editing process. As a result, the common group is not quite the same in each of the calibrations.
2. Although the tests were designed to represent tests that were easy, hard, and matched relative to the average ability of the common group, the results did not reflect this. Due to the anchoring process, all three test calibrations yielded a mean item difficulty of zero. For putting the three tests on a common ability scale (i.e., test equating), see Appendix C.
3. Within each calibration, examinees with the same raw test score obtained the same estimated ability. However, a given raw score will not yield the same estimated ability across the three calibrations.
4. Even though the same group of examinees was administered all three tests, the means and the standard deviations of their ability estimates were different for each calibration. This can be attributed to a number of causes. The primary reason is that, due to the anchoring process, the value of the mean estimated abilities is expressed relative to the mean item difficulty of the test. Thus, the mean difficulty of the easy test should result in a positive mean ability. The mean ability on the

7.7 Things to Notice

hard test should have a negative value. The mean ability on the matched test should be near zero. The changing group membership also accounts for some of the differences, particularly when the group was small to start with. Finally, the overall amount of information is rather small in all three test information functions. Thus, the ability level of none of the examinees is being estimated very precisely. As a result, the ability estimate for a given examinee is not necessarily very close to his or her true ability.

5. The anchoring procedure set the mean item difficulty equal to zero, and thus the midpoint of the ability scale to zero. A direct consequence of this is that the mid true score for all three test characteristic curves occurs at an ability level of zero (see Fig. 7.1). The similarity in the shapes of the curves for the first two data sets was due to the item difficulties being distributed in an approximately symmetrical manner around the zero point. The fact that all the items had the same value of the discrimination parameter (1.0) makes the slopes of the first two curves similar. The curve for data set 3 falls below those for data sets 1 and 2, as it was based on only eight items. However, its general shape is similar to the previous two curves and its mid true score occurred at an ability level of zero.
6. Although the test information functions were similar, there were some important differences (see Fig. 7.2). The curve for the matched test had the same general level as that for the easy test. The curve was a bit flatter, indicating this test maintained its level of precision over a wide range. The test information function for the hard test had a slightly smaller amount of information at its midpoint. Thus, it had a bit less precision at this point. The curve was a bit lower than the other two, indicating it did not hold the same precision over the usual range of ability.

Chapter 8
Specifying the Characteristics of a Test

8.1 Introduction

During this transitional period in testing practices, many tests have been designed and constructed using classical test theory principles but have been analyzed via item response theory procedures. This lack of congruence between the construction and analysis procedures has resulted in the full power of item response theory not being exploited. In order to obtain the many advantages of item response theory, tests should be designed, constructed, analyzed, and interpreted within the framework of the theory. Consequently, the goal of this chapter is to provide the reader with experience in the technical aspects of test construction within the framework of item response theory.

Persons functioning in the role of test constructors do so in a wide variety of settings. They develop tests for commercial testing companies, governmental agencies, and school districts. In addition, teachers at all classroom levels develop tests to measure achievement. In all of these settings, the test-construction process is usually based upon having a collection of items from which to select those to be included in a particular test. Such collections of items are known as item pools. Items are selected from such pools on the basis of both their content and their technical characteristics; that is, their item parameter values. Under item response theory, a well-defined set of procedures is used to establish and maintain such item pools. A special name, item banking, has been given to these procedures. The basic goal is to have an item pool in which the values of the item parameters are expressed in a known ability scale metric. If this is done, it is possible to select items from the item pool and determine the major technical characteristics of a test before it is administered to a group of examinees. If the test characteristics do not meet the design goals, selected items can be replaced by other items from the item pool until the desired characteristics are obtained. Considerable time and money are saved that would ordinarily be devoted to piloting the test.

In order to build an item pool, it is necessary first to define the latent trait the items are to measure, write items to measure this trait, and pilot-test the items to weed out poor items. After some time, a set of items measuring the latent trait of interest is available. This large set of items is then administered to a large group of examinees. An item characteristic curve model is selected, the examinees' item response data are analyzed via the Birnbaum paradigm, and the test is calibrated. The ability scale resulting from this calibration is considered to be the baseline metric of the item pool. From a test construction point of view, we now have a set of items whose item parameter values are known, and in technical terms, a "precalibrated item pool" exists.

8.2 Developing a Test from a Precalibrated Item Pool

Since the items in the precalibrated item pool measure a specific latent trait, tests constructed from it will also measure this trait. While this may seem a bit odd, there are a number of reasons for wanting additional tests to measure the same trait. For example, alternate forms are routinely needed to maintain test security and special versions of the test can be used to award scholarships. In such cases, items would be selected from the item pool on the basis of their content and their technical characteristics to meet the particular testing goals.

The advantage of having a precalibrated item pool is that the parameter values of the items included in the test can be used to compute the test characteristic curve and the test information function before the test is administered. This is possible because neither of these curves depends upon the distribution of examinee ability scores over the ability scale. Thus, both curves can be obtained once the values of the item parameters are available. Given these two curves, the test constructor has a very good idea of how the test will perform before it is given to a group of examinees. In addition, when the test has been administered and calibrated, test equating procedures can be used to express the ability estimates of the new group of examinees in the metric of the item pool.

8.3 Some Typical Testing Goals

In order to make the computer exercises meaningful to you, several types of testing goals are defined below. These will then serve as the basis for specific types of tests you will create.

1. Screening tests.

    Tests used for screening purposes have the capability to distinguish rather sharply between examinees whose abilities are just below a given ability level and those who are at or above that level. Such tests are used to assign scholarships and to assign students to specific instructional programs such as remediation or advanced placement.

2. Broad-range tests.

These tests are used to measure ability over a wide range of underlying ability scale. The primary purpose is to be able to make a statement about an examinee's ability and to make comparisons among examinees. Tests measuring reading or mathematics are typically broad-range tests.

3. Peaked tests.

Such tests are designed to measure ability quite well in a region of the ability scale where most of the examinees' abilities will be located, and less well outside this region. When one deliberately creates a peaked test, it is to measure ability well in a range of ability that is wider than that of a screening test, but not as wide as that of a broad-range test.

8.4 Computer Session

The purpose of this session is to assist you in developing the capability to select items from a precalibrated item pool to meet a specific testing goal. You will set the parameter values for the items of a small test in order to meet one of the three testing goals given above. Then the test characteristic curve and the test information function will be shown on the screen and you can determine if the testing goal was met. If not, a new set of item parameters can be selected and the resultant curves obtained. With a bit of practice, you should become proficient at establishing tests having technical characteristics consistent with the design goals.

8.4.1 Some Ground Rules

1. It is assumed that the items would be selected on the basis of content as well as parameter values. For present purposes, the actual content of the items need not be shown.
2. No two items in the item pool possess exactly the same combination of item parameter values.
3. The item parameter values are subject to the following constraints:

$$-3.0 \leq b \leq +3.0$$
$$0.5 \leq a \leq +2.0$$
$$0 \leq c \leq .35$$

The values of the discrimination parameter have been restricted to reflect the range of values usually seen in well-maintained item pools.

8.4.2 Procedures for an Example Case

You are to construct a ten-item screening test that will separate examinees into two groups: Those who need remedial instruction and those who don't, on the ability measured by the items in the item pool. Students whose ability falls below a value of -1.0. will receive the instruction.

1. The number of items in the test is $J = 10$. The number will be determined by the length of the vector of item difficulty parameters.
2. The item characteristic curve model is the two-parameter model. The number of the vectors of item parameters will determine the model.
3. Set the following item parameter values for a first test:

Item	Difficulty b	Discrimination a
1	−1.8	1.2
2	−1.6	1.4
3	−1.4	1.1
4	−1.2	1.3
5	−1.0	1.5
6	−0.8	1.0
7	−0.6	1.4
8	−0.4	1.2
9	−0.2	1.1
10	0.0	1.3

    The logic underlying these choices was one of centering the difficulties on the cut off level of -1.0 and using moderate values of item discrimination. The R command lines for the first set of item parameter vectors are as follows:

    ```
    > b1 <- c(-1.8, -1.6, -1.4, -1.2, -1.0,
         -0.8, -0.6, -0.4, -0.2, 0.0)
    > a1 <- c(1.2, 1.4, 1.1, 1.3, 1.5, 1.0, 1.4, 1.2, 1.1, 1.3)
    ```

4. Study the table of item parameters for a moment. If you need to change a value in the R command lines, you can click the up or down arrow key to show the command lines you have typed in earlier will appear. You can enter a new value without typing in all other values.
5. When you are satisfied with the parameter values, you can type in the R command lines to construct the true score function and a graph of the test characteristic curve. You may use the function `tcc`.
6. By executing the R command lines, the test characteristic curve shown in Fig. 8.1 will appear on the screen in the graphics window.
7. When the test characteristic curve appears on the screen, make note of the ability level at which the mid true score occurs. Also note the slope of the curve at that ability level.

8.4 Computer Session

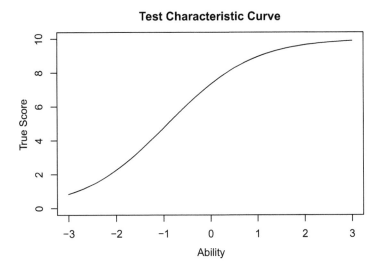

Fig. 8.1 Test characteristic curve for the example

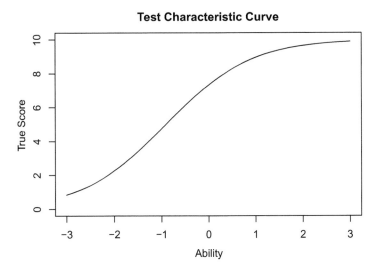

Fig. 8.2 Test information function for the example

8. You can type in the R command lines to construct the test information function and to display it in the graphics window. You may use the function `tif`.
9. By typing in and executing the R command lines, the test information function will appear on the screen. Note the maximum amount of information and the ability level at which it occurred. The function is shown in Fig. 8.2.
10. Assuming that you are using the two R functions, `tcc` and `tif`, you can construct the two graphs in the same graphics window. After executing the

functions and entering the vectors of item parameters, use the following R command lines:

```
> par(mfrow = c(2,1))
> tcc(b1, a1)
> tif(b1, a1)
```

11. On the top of the graphics window, the test characteristic curve will appear. On the bottom of the graphics window, the test information function will show up. Thus, you can contrast two graphs to study their relationship. When you are ready to return to one plot per figure setting, you may use:

```
> par(mfrow = c(1,1))
```

12. The design goal was to specify the items of a screening test that would function at an ability level of -1.0. In general, this goal has been met. The mid true score corresponded to an ability level of -1.0. The test characteristic curve was not particularly steep at the cut off level indicating that the test lacked discrimination. The peak of the information function occurred at an ability level of -1.0 but the maximum was a bit small. The results suggest that the test was properly positioned on the ability scale but that a better set of items could be found. The following changes would improve the test's characteristics: first, cluster the values of the item difficulty parameters nearer the cut off level; second, use larger values of the item discrimination parameters. These two changes should steepen the test characteristic curve and increase the maximum amount of information at the ability level of -1.0.

13. Now using the same number of items and the same item characteristic curve model, a new, second test will be created. The item parameters are as follows:

```
> b2 <- c(-1.1, -1.0, -1.1, -1.2, -1.0,
    -0.8, -0.9, -1.0, -0.9, -1.0)
> a2 <- c(1.9, 1.7, 1.8, 1.6, 1.9, 1.8, 1.9, 1.9, 1.7, 1.6)
```

Now try to obtain the graph of the test characteristic curve for the first test and then obtain that of the second test using the R command lined. You can plot them onto the same graph or two separate graphs can be obtained. For example, you can plot the two test characteristic curves on the same graph using:

```
> par(mfrow = c(1,1))
> tcc(b1, a1)
> par(new=T)
> tcc(b2, a2)
```

When the test characteristic curves appear sequentially, then compare them. Determine if you have increased the slope of the curve at the ability level -1.0 using the second test.

14. The two graphs of the test information functions can be plotted on the top and bottom of the graphics window using the following R command:

    ```
    > par(mfrow = c(2,1))
    > tif(b1, a1)
    > tif(b2, a2)
    ```

15. When the test information functions appear sequentially, then compare them. Determine if the maximum amount of information is larger for the second test than it was for the first test at an ability level of -1.0.
16. If all went well the new set of test items should have improved the technical characteristics of the test as reflected in the test characteristic curve and the test information function.

8.5 Exercises

In each of the following exercises, establish a set of item parameters. After you have seen the test characteristic curve and the test information function, use the R command lines and to construct a new set of vectors of item parameters by changing selected item parameter values. Also overlay the new curves on the previous curves. These procedures will allow you to see the impact of the changes. Repeat this process until you feel that you have achieved the test specification goal.

1. Construct a ten-item screening test to function at an ability level of $+0.75$ using a Rasch model.
2. Construct a broad-range test under the three-parameter model that will have a horizontal test information function over the ability range of -1.0 to $+1.0$.
3. Construct a test having a test characteristic curve with a rather small slope and a test information function that has a moderately rounded appearance. Use either the two- or three-parameter model.
4. Construct a test that will have a nearly linear test characteristic curve whose mid-true score occurs at an ability level of zero. Use the Rasch model.
5. Repeat the previous problem using the three-parameter model.
6. Construct a test that will have a horizontal test information function over the ability range of -2.0 to $+2.0$ having a maximum amount of information of 2.5.
7. Use the computer session to experiment with different combinations of testing goals, item characteristic curve models, and numbers of items. The goal is to be able to obtain test characteristic curves and test information functions that are optimal for the testing goals. It will be helpful to use the editing feature of the R (i.e., arrow keys) to change specific item parameter values rather than re-enter a complete set of item parameter values for each trial.

8.6 Things to Notice

1. Screening tests.

   (a) The desired test characteristic curve has the mid true score at the specified cut off ability level. The curve should be as steep as possible at that ability level.

   (b) The test information function should be peaked with its maximum at the cut off ability level.

   (c) The values of the item difficulty parameters should be clustered as closely as possible around the cut off ability of interest. The optimal case is where all values of the item difficulty parameters are at the cut off point and the values of the item discrimination parameters are large. However, this is unrealistic because an item pool rarely contains enough items with common difficulty values. If a choice among items must be made, select items that yield the maximum amount of information at the cut off point.

2. Broad-range tests.

   (a) The desired test characteristic curve has its mid true score at an ability level corresponding to the mid-point of the range of ability of interest. Most often this is an ability level of zero. The test characteristic curve should be linear for most of its range.

   (b) The desired test information function is horizontal over the widest possible range. The maximum amount of information should be as large as possible.

   (c) The values of the item difficulty parameters should be spread uniformly over the ability scale and as widely as practical. There is a conflict between the goals of a maximum amount of information and a horizontal test information function. To achieve a horizontal test information function, items with low to moderate item discrimination parameters that have a U-shaped distribution of item difficulty parameters are needed. However, such items yield a rather low general amount of information and the overall precision will be low.

3. Peaked tests.

   (a) The desired test characteristic curve has its mid true score at an ability level in the middle of the ability range of interest. The curve should have a moderate slope at that ability level.

   (b) The desired test information function should have its maximum at the same ability level as the mid true score of the test characteristic curve. The test information function should be rounded in appearance over the ability range of most interest.

   (c) The item difficulty parameters should be clustered around the mid-point of the ability range of interest, but not as tightly as in the case of a screening test. The values of the discrimination parameters should be as large as practical. Items whose values of the item difficulty parameters are within the ability range of interest should have larger values of the item discrimination parameters than items whose values of the item difficulty parameters are outside this range.

4. Role of item characteristic curve models.
   (a) Due to the value of the discrimination parameters being fixed at 1.0, the Rasch model has a limit placed upon the maximum amount of information that can be obtained. The maximum amount of item information is 0.25 since $P_j(\theta)Q_j(\theta) = 0.25$ when $P_j(\theta) = 0.5$. Thus, the theoretical maximum amount of information for a test under the Rasch model is 0.25 times the number of items.
   (b) Due to the presence of the guessing parameter, the three-parameter model will yield a more linear test characteristic curve and a test information function with a lower general level than under the two-parameter model with the same set of item difficulty and discrimination parameters. The information function under the two-parameter model is the upper bound for the information function under the three-parameter model when the values of b and a are the same.
   (c) For test specification purposes, the authors prefer the two-parameter model.
5. Role of the number of items.
   (a) Increasing the number of items has little impact upon the general form of the test characteristic curve if the distribution of the sets of item parameters remains the same.
   (b) Increasing the number of items in a test has a significant impact upon the general level of the test information function. The optimal situation is a large number of items having high values of the item discrimination parameters and a distribution of item difficulty parameters consistent with the testing goals.
   (c) The manner in which the values of the item parameters are paired is an important consideration. For example, choosing a high value of the item discrimination parameter for an item whose item difficulty parameter is not of interest does little in terms of the test information function or the slope of the test characteristic curve. Thus, the test constructor must visualize both what the item characteristic curve and the item information function looks like in order to ascertain its contribution to both the test characteristic curve and the test information function.

Appendix A
R Introduction

A.1 R Installation

R is a programming environment for data analysis and graphics (Venables et al. 2009; see Appendix A references in the end). R can also be seen as an implementation of the S language (see Becker and Chambers 1984, 1985; Becker et al. 1988; Chambers 1998; Chambers and Hastie 1993; Spector 1994). In an everyday ordinary vernacular, R is simply a free computer program for statistics.

The main website for R is:

http://www.r-project.org

The binary executable files for installing R for Linux, Mac OS X, and Windows can be obtained via the Comprehensive R Archive Network (CRAN) for which the link is provided in the main website. Note that the main website contains other relevant information including the R manuals and a list of the books related to R. The base distribution package of the 64-bit R for Windows is the program discussed here. The main uniform resource locator (URL) of the R base distribution is:

https://cran.rstudio.com/bin/windows/base/

By clicking the download link on the web page, the executable file will be saved in the default download directory of your computer. It took 11 s to download the executable code using the second author's office computer. By double clicking the file and specifying options to choose (e.g., English as the setup language, clicking the Next button seven times without changing any default specifications, and clicking the Finish button), you can install R onto your computer.

If properly installed, the R icon can be found in the desktop of the Windows screen. To invoke R, double click the R icon. When invoked, the computer will show the R console window that looks like the one presented in Fig. A.1.

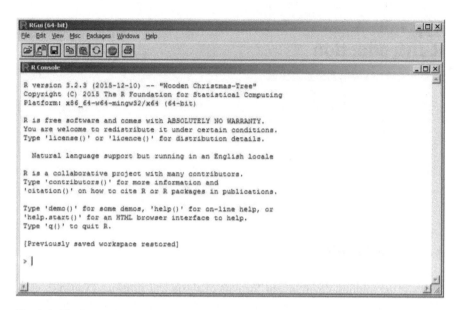

Fig. A.1 The R console window

To quit R, type in

> q()

and press the enter key in the R console window. R shows a question window asking if you want to "Save work space image?" Only if you have variables and objects to be continuously used in the next session when you invoke R again, you will click the Yes button. Otherwise, you may click the No button to exit from the current R session.

A.2 R Basics

A.2.1 Scalar Arithmetic

R has the usual arithmetic operators:

- `+` for addition
- `-` for subtraction
- `*` for multiplication
- `/` for division
- `^` for exponentiation (i.e., raising to a power)
- `%/%` for integer division
- `%%` for remainder from integer division

A.2 R Basics

There is also an arithmetic operator − for unary minus that is applicable to one operand (i.e., making a negative value; cf. + can also be used as unary plus). For example, the following command line (i.e., expression) yields, based on the order of operations (i.e., ^ first, * / second, and + − last, from left to right if the orders are the same), the answer shown in the next line:

```
> 1 + 2 - 3 * 4 / 5 ^ 6
[1] 2.999232
```

The number in the brackets (e.g., [1]) indicates the order of elements in the result. Because we are getting a scalar value, only one number is shown after such a bracketed number. This will be handy where we operate with vectors instead of scalars. The order of operations can be changed with the use of parentheses. For example:

```
> (1 + 2 - 3) * 4 / 5 ^ 6
[1] 0
> 1 + 2 - (3 * 4 / 5) ^ 6
[1] -188.103
> 1 + (2 - 3) * 4 / 5 ^ 6
[1] 0.999744
> 1 + (2 - 3) * (4 / 5) ^ 6
[1] 0.737856
```

The portions of the command line beginning with the pound symbol, #, to the end of the line before pressing the enter key will be treated as a comment. For example:

```
> 6 + 5 - 4 * 3 / 2 ^ 1 # the answer is [Enter]
[1] 5
```

Comments can be typed (and ignored) anywhere in the R expressions. Comments can be very informative explaining the command line or expression to be carried out by R.

The default number of decimal places is 7. It can be changed with the options function with the digits argument for which the valid values are 0–22. It should be noted that there may exist rounding errors when a very larger number of decimal places, say 22, is employed. For example, 1/3 will yield 0.3333333 in the default setting, but the following rounding error can occur when a higher precision is requested:

```
> options(digits=22)
> 1 / 3
[1] 0.3333333333333333148296
```

With the same options function in effect, the mathematical constant π can be obtained using R both directly with the pi command and indirectly with the arc tangent function:

```
> pi
[1] 3.141592653589793115998
> 4 * atan(1)
[1] 3.141592653589793115998
```

Note that the number obtained may not be consistent with such a number reported in other sources (e.g., $\pi = 3.141592653589793238462643$ from Abramowitz and Stegun 1964, p. 3). In many practical settings, it will be sufficient to use the default setting of the number of decimal places, that is:

```
> options(digits=7)
```

Other mathematical constants can be obtained using R functions. In fact, nearly all of the common mathematical functions are available in R with arguments in parentheses (i.e., parenthetical arguments). For example, mathematical functions include:

abs	absolute value
exp	exponential (e to a power)
gamma	gamma function
lgamma	log of gamma function
log	logarithm
log10	logarithm of base 10
sign	signum function
sqrt	square root
floor	largest integer, less than or equal to
ceiling	smallest integer, greater than or equal to
trunc	truncation to the nearest integer
factorial	factorial
lfactorial	log of factorial

A full range of logical operators can be used in R:

>	greater than
<	less than
>=	greater than or equal to
<=	less than or equal to
==	equality
!=	non-equality
&	elementwise and
|	elementwise or
&&	control and
||	control or
!	unary not

A.2 R Basics

Trigonometric functions available in R include the followings:

```
cos     cosine
sin     sine
tan     tangent
acos    arc cosine
asin    arc sine
atan    arc tangent
cosh    hyperbolic cosine
sinh    hyperbolic sine
tanh    hyperbolic tangent
acosh   arc hyperbolic cosine
asinh   arc hyperbolic sine
atanh   arc hyperbolic tangent
```

In the trigonometric functions, the arguments are in radians instead of degrees. For example:

```
> sin(pi / 6)
[1] 0.5
> pi / 6
[1] 0.5235988
> sin(0.5235988)
[1] 0.5
> sin((30 / 180) * pi)
[1] 0.5
```

When an R command is not grammatically complete but gets the enter key, the prompt will be changed to a plus sign indicating some additional input is necessary. After completing the grammatically correct command line and pressing the enter key, the command will be executed. For example:

```
> sin(pi / 6
+ )
[1] 0.5
```

A scalar value or the result from arithmetic operators can be saved as a variable with the assignment function <-. The value can be listed by typing in the variable name:

```
> a <- sin(pi / 6)
> b <- cos(pi / 6)
> c <- sqrt(a ^ 2 + b ^ 2)
> c
[1] 1
```

Multiple command lines can be combined by separating them with semicolons. Spaces are mostly optional in the R commands, but readability will be enhanced when proper spacing is employed. For example:

```
> a<-sin(pi/6);b<-cos(pi/6);c<-sqrt(a^2+b^2);c
[1] 1
```

R can handle operations of complex numbers that have real parts and imaginary parts albeit not really useful in applied statistical procedures:

```
> x <- 4 + 2i
> Re(x)
[1] 4
> Im(x)
[1] 2
> y <- 4 - 2i
> x + y
[1] 8+0i
> x * y
[1] 20+0i
```

A.2.2 Vector Arithmetic

Here, a vector is a single entity consisting of an ordered collection of numbers. After defining a vector in R, arithmetic operations and functional operations can also be performed with the vector. To set up a vector named x consisting of four numbers, namely, 1, 2, 3, and 2, we may use the assign function with the combine/concatenate function for which the elements are separated with commas:

```
> x <- c(1, 2, 3, 2)
```

After defining the vector (i.e., a variable in a statistical sense), the elements of the vector can be listed by typing in the name of the vector:

```
> x
[1] 1 2 3 2
```

The edit function used with the assign function to the same object can be used to modify and save the elements of the vector. For example,

```
> x <- c(1, 2, 3, 2)
> x <- edit(x)
```

will open the R editor window and allow to change the elements. After finishing editing, a new vector can be constructed either by clicking the File-Save option in the main R console window or by clicking the close window icon and then pressing the Yes button in the question dialogue box. Note that the edit function is not limited

A.2 R Basics

to the vectors but applicable to other objects including a blank vector (n.b., it may bring up the last object edited by the R editor instead of the blank page):

```
> x <- edit()
```

Also note that depending upon the types of objects used in the edit function, the R data editor window will be used (e.g., for matrices and data.frame) instead of the R editor window.

Functions for simple statistics for a vector are available in R:

min	smallest value
max	largest value
range	minimum and maximum
mean	arithmetic average
var	variance
sd	standard deviation
sum	arithmetic sum
prod	product of elements
length	number of elements
median	50th percentile
quantile	quantiles
cumsum	cumulative sum
diff	first difference
table	frequency table or crosstabulation
summary	five number summary or frequencies

In addition, after defining two vectors, the following statistical functions are available in R:

cor	correlation
cov	covariance

For example:

```
> x <- c(1, 2, 3, 2)
> y <- c(1, 3, 2, 2)
> cor(x, y)
[1] 0.5
> cov(x, y)
[1] 0.3333333
```

Sorting or rearranging of the vector in the ascending or increasing order and in the descending or decreasing order can be performed using the sort function, for example:

```
> x <- c(1, 2, 3, 2)
> sort(x)
[1] 1 2 2 3
> sort(x, decreasing=T)
[1] 3 2 2 1
```

A subset of vector can be created by using the order subscripts and their operations in brackets, for example:

```
> x <- c(1, 2, 3, 2)
> x[1]
[1] 1
> x[2 : 4]
[1] 2 3 2
> x[-3]
[1] 1 2 2
> x[x < 3]
[1] 1 2 2
> x[x > 2]
[1] 3
```

Note that the vector can be replaced with the assignment function, for example:

```
> x <- x[-3]; x
[1] 1 2 2
```

The elements in a vector are not limited to numbers. A logical vector and a character vector can be used.

```
> x <- c(1, 2, 3, 2)
> x < 3
[1]   TRUE   TRUE  FALSE    TRUE
> labels <- c("red", "white", "blue", "white"); labels
[1] "red"   "white" "blue"  "white"
> names(x) <- labels; x
  red white  blue white
    1     2     3     2
> names(x) <- NULL; x
[1] 1 2 3 2
```

Vectors can be generated and converted to different types using functions in R:

numeric	a vector of zeros with the length of the argument
character	a vector of blank characters of argument length
logical	a vector of FALSE of argument length
seq	argument 1 to argument 2 with the increment of argument 3
1 : 4	numbers equivalent to seq(1, 4, 1)
rep	replicate argument1 as many times as argument 2
as.numeric	conversion to numeric
as.character	conversion to string-type
as.logical	conversion to logical
factor	creating factor from vector

A.2 R Basics

For example, the followings are very useful ways to construct a sequence of nicely patterned elements:

```
> x <- 1 : 4; x
[1] 1 2 3 4
> x <- seq(1, 4, 1); x
[1] 1 2 3 4
> x <- seq(1, 2, 0.2); x
[1] 1.0 1.2 1.4 1.6 1.8 2.0
> x <- rep(1, 4); x
[1] 1 1 1 1
> x <- c(rep(1,4), rep(2,2)); x
[1] 1 1 1 1 2 2
```

A.2.3 Matrices and Matrix Functions

An array is a collection of data which can be indexed by one or more subscripts. The vectors discussed in the previous section can be seen as one-dimensional arrays. Each element in a vector can be referred to as the name with the subscript enclosed in brackets (e.g., x[1]). Two-dimensional arrays are generally referred to as matrices. The matrix function is used to create a matrix. For example, a matrix with ones in the first column and four observations in the second column can be defined and listed subsequently by:

```
> X <- matrix(c(1, 1, 1, 1, 1, 2, 3, 2), nrow=4); X
     [,1] [,2]
[1,]    1    1
[2,]    1    2
[3,]    1    3
[4,]    1    2
```

The R commands as well as the names of objects and variables are case sensitive. The objects X and x, for example, are not the same unless these are defined to be equivalent. The command line of the above matrix is equivalent to:

```
> X <- matrix(c(1, 1, 1, 1, 1, 2, 3, 2), ncol=2)
> X <- matrix(c(1, 1, 1, 1, 1, 2, 3, 2), nrow=4, ncol=2)
> X <- matrix(c(1, 1, 1, 2, 1, 3, 1, 2), nrow=4, byrow=T)
> X <- matrix(c(1, 1, 1, 2, 1, 3, 1, 2), ncol=2, byrow=T)
> X <- matrix(c(1,1,1,2,1,3,1,2), nrow=4, ncol=2, byrow=T)
```

Elements in a matrix can be referred to as the name with the row and column subscripts enclosed in brackets. For example, with the same matrix defined earlier:

```
> X[2,2]
[1] 2
```

```
> X[,2]
[1] 1 2 3 2
> X[2,]
[1] 1 2
> X[1 : 2,]
     [,1] [,2]
[1,]   1    1
[2,]   1    2
```

After defining two or more vectors of the same length (i.e., the same number of elements), a matrix can be constructed by the cbind function:

```
> u <- c(1, 1, 1, 1)
> x <- c(1, 2, 3, 2)
> X <- cbind(u, x); X
     u x
[1,] 1 1
[2,] 1 2
[3,] 1 3
[4,] 1 2
```

It can be noticed that the default column names in the listing of the matrix are replaced with the names of the vectors. The equivalent matrix function is:

```
> X <- matrix(c(1, 1, 1, 1, 1, 2, 3, 2), ncol=2,
    dimnames=list(c(),c("u","x")))
```

Also the row and column names can be specified with the functions of rowname and colnames, respectively:

```
> X <- matrix(c(1, 1, 1, 1, 1, 2, 3, 2), ncol=2)
> colnames(X) <- c("u", "x")
> rownames(X) <- c()
```

After defining vectors of the same length in row wise, a matrix can be constructed by the rbind function:

```
> r1 <- c(1, 1); r2 <- c(1, 2); r3 <- c(1, 3); r4 <- c(1, 2)
> X <- rbind(r1, r2, r3, r4); X
   [,1] [,2]
r1   1    1
r2   1    2
r3   1    3
r4   1    2
> rownamess(X) <- c()
```

Note that the row names can be replaced with the default names with the rowname function as shown in the last line.

A.2 R Basics

A matrix can also be constructed with the array function, although the array is not limited to be two-dimensional. For example:

```
> X <- array(c(1, 1, 1, 1, 1, 2, 3, 2), dim=c(4,2)); X
     [,1] [,2]
[1,]   1    1
[2,]   1    2
[3,]   1    3
[4,]   1    2
```

Once a matrix is defined, the dimension, the number of rows, and the number of columns of the matrix can be obtained with the following functions:

```
> dim(X)
[1] 4 2
> nrow(X)
[1] 4
> ncol(X)
[1] 2
```

The followings are some matrix functions:

chol	Cholesky decomposition
crossprod	matrix crossproduct
det	determinant
diag	to create or extract diagonal values
eigen	eigenvalues and eigenvectors
outer	outer product of two vectors
scale	to scale the columns of a matrix
solve	inversion or to solve system of linear equations
svd	singular value decomposition
qr	qr orthogonalization
t	to transpose

Based on the usual conforming conditions with scalars and matrices, the element-wise addition, subtraction, multiplication, and division can be performed. Matrix multiplication is done with the operator:

%*%	matrix multiplication

The following is an example to obtain the estimates of an intercept and a slope from a simple regression model using the matrix functions and operators:

```
> X <- array(c(1, 1, 1, 1, 1, 2, 3, 2), dim=c(4,2))
> colnames(X) <- c("u", "x")
> y <- c(1, 3, 2, 2)
> solve(t(X) %*% X) %*% t(X) %*% y
    [,1]
u    1.0
```

```
x   0.5
> betahat <- solve(crossprod(X, X)) %*% t(X) %*% y;
> rownames(betahat) <- c("a", "b"); betahat
   [,1]
a  1.0
b  0.5
> ypredict <- X %*% betahat; ypredict
      [,1]
[1,]  1.5
[2,]  2.0
[3,]  2.5
[4,]  2.0
> yhat <- ypredict[,1]; yhat
[1] 1.5 2.0 2.5 2.0
> residual <- y - yhat; residual
[1] -0.5  1.0 -0.5  0.0
```

The results from regression analysis in general will be obtained not from the matrix or vector operations but from the R functions for statistical modeling. Hence, the above command line illustrations are only for the demonstration and instructional purpose.

The eigenvalues and eigenvectors from a matrix can be obtained, for example:

```
> R <- matrix(c(1, .5, 0, .5, 1, .5, 0, .5, 1), nrow=3)
> R
     [,1] [,2] [,3]
[1,]  1.0  0.5  0.0
[2,]  0.5  1.0  0.5
[3,]  0.0  0.5  1.0
> eigen(R)
$values
[1] 1.7071068 1.0000000 0.2928932

$vectors
          [,1]           [,2]         [,3]
[1,] 0.5000000 -7.071068e-01 -0.5000000
[2,] 0.7071068 -1.099065e-15  0.7071068
[3,] 0.5000000  7.071068e-01 -0.5000000
```

A.2.4 Data Frame

A data frame is a two-dimensional array of observations in rows and variables in columns. Functions such as dim, dimnames, nrow, and ncol will work on data frames. The attach function for data frames allow that variables contained in the

A.2 R Basics

data frame can be easily accessed through the variable names. Data frames can be constructed by the data.frame function:

```
> x <- c(1, 2, 3, 2)
> y <- c(1, 3, 2, 2)
> X <- data.frame(x, y); X
  x y
1 1 1
2 2 3
3 3 2
4 2 2
```

Variables can be extracted from the data frame or directly referred by declaring the data frame name and the variable name separated with a dollar sign. For example, the x vector can be listed with the following commands assuming that the data frame has been declared as in the earlier command lines:

```
> X[,1]
> X$x
```

The names function displays the variable names in the data frame:

```
> names(X)
[1] "x" "y"
```

A new variable can be appended to an existing data frame with a dollar sign and a variable name using the combine function. A variable can be a string variable, for example:

```
> X$z <- c("a", "b", "c", "d"); X
  x y z
1 1 1 a
2 2 3 b
3 3 2 c
4 2 2 d
```

A variable can be removed or portions of the variables can be selected as in the following command lines for the previous data frame X with the three variables:

```
> X <- X[,-3]
> X <- X[,1 : 2]
```

These yield the same data frame X with only x and y variables.

The edit function opens the data editor window and allows to edit the data frame with the spreadsheet-looking data editor. The values of the variables as well as the variable names can be modified. The data frame can be saved by the clicking of the close window icon, that is, the exit button positioned in the top, right corner of the data editor window's title bar.

It is also possible to construct a data frame by opening up a blank data frame using the edit function and then entering the necessary values and variable names. The command line looks:

```
> X <- edit(data.frame())
```

Although not terribly useful, the data frame can be saved as an R system file (cf. that cannot be meaningfully read by other text editors; cf. saveRDS, readRDS) and retrived or loaded to R:

```
> save(X, file="X.RData")
> load("X.RData")
```

The file will be saved in the default, current working directory specified in the installation process, for example, the exact location of the file can be:

C:\Users\shkim\Documents\X.RData

A data frame can be saved as a file that can be opened with other editor-type programs as:

```
> write.table(X, file="X.txt", sep=" ")
```

The current working directory where the data frame file is to be stored can be found with the command line

```
> getwd()
```

and the directory can be changed to the usual root directory with either of the following two command lines:

```
> setwd("C:\\")
> setwd("C:/")
```

After loading the file, the variables contained in a data frame can be directly accessed by declaring the attach function:

```
> attach(X)
```

A data frame can be removed from the current session with the detach function, for example:

```
> detach(X)
```

If there is an object defined with the same variable name in the attached object, then due to a hierarchical nature of searching objects in the R workspace the attach function may not bring up the variable contained in the data frame. Care should be exercised when the attach function is employed.

A list of currently available objects can be found by the ls function:

```
> ls()
```

The objects can be removed by the rm function, for example:

```
> x <- c(1, 2, 3, 2)
> ls()
[1] "x"
> rm(x)
```

The entire workspace will be cleared by:

```
> rm(list=ls())
```

A.2 R Basics

A data file written in ASCII or a text file can be read into R using the read.fwf (i.e., fixed width format) function. Suppose there is a text file named crux.txt with three variables in each record in the main directory of the computer:

```
acrux 11
mimosa23
gacrux32
palida22
```

The lengths based on the numbers of the columns of the three variables are 6, 1, and 1, respectively. The file can be read in two trivially different ways employing different directory symbols in the file definitions:

```
> crux <- read.fwf("c:\\crux.txt", width=c(6,1,1))
> crux
      V1 V2 V3
1 acrux  1  1
2 mimosa 2  3
3 gacrux 3  2
4 palida 2  2
> crux <- read.fwf("c:/crux.txt", width=c(6,1,1))
> crux
      V1 V2 V3
1 acrux  1  1
2 mimosa 2  3
3 gacrux 3  2
4 palida 2  2
```

Variables in a text data file can be separated with either blank spaces or commas. Suppose there is a text file named cruxb.txt with three variables separated with blank spaces in the main directory of the computer:

```
acrux 1 1
mimosa 2 3
gacrux 3 2
palida 2 2
```

The file can be read in with the following command line:

```
> crux <- read.table("c:/cruxb.txt", sep=" ")
```

Suppose there is a text file named cruxc.txt with three variables separated with commas in the main directory of the computer:

```
acrux,1,1
mimosa,2,3
gacrux,3,2
palida,2,2
```

Such a file can be read in with the following command line:

```
> crux <- read.table("c:/cruxc.txt", sep=",")
```

Note that the default variable names can be modified with the colnames function, for example, after reading in the crux file:

```
> colnames(crux) <- c("name", "x", "y")
> crux
    name x y
1  acrux 1 1
2 mimosa 2 3
3 gacrux 3 2
4 palida 2 2
```

It should be noted that R is not a good program to handle large, practical data sets. There are better computer programs to use for typical applied statistical analyses (e.g., SAS, SPSS, etc.). You will get what you paid for from these other programs.

A.2.5 Probability Distributions

Functions for probability distributions can be classified into four different types: (1) density (i.e., probability density function), (2) probability (i.e., cumulative density function), (3) quantile, and (4) random sample. A prefix as the first starting letter is attached to the name of a distribution to construct the four types of functions. For a normal distribution, the distribution name is abbreviated as norm. Hence, dnorm, pnorm, qnorm, and rnorm are used to obtain the respective functions with some specified arguments. Because the normal distribution requires the two parameters, that is, mean and sd as arguments, the two parameters should be specified as arguments unless the default standard normal distribution is in play. For the standard normal distribution, some examples are:

```
> dnorm(0)
[1] 0.3989423
> pnorm(-1.96)
[1] 0.0249979
> qnorm(0.025)
[1] -1.959964
> rnorm(3)
[1] -0.5514404  0.0889554  0.3656794
```

Note that a different normal distribution can be used to perform similar tasks. If the mean of the normal distribution is 100 and the standard deviation of the normal distribution is 15, then the above operations can be similarly done via:

```
> dnorm(100, 100, 15)
[1] 0.02659615
```

A.2 R Basics

```
> pnorm((-1.96*15+100), 100, 15)
[1] 0.0249979
> qnorm(0.025, 100, 15)
[1] 70.60054
> rnorm(3, 100, 15)
[1] 108.92949  83.03011 117.58789
```

The random sampling is performed by resetting the current random number seed saved in the .Random.seed vector. In case of generating the same sequence of the randomly sampled values, the following can be performed:

```
> oldseed <- .Random.seed
> rnorm(3, 100, 15)
[1] 107.21030  79.72381 101.61939
> rnorm(3, 100, 15)
[1] 85.63351 89.31402 94.36237
> .Random.seed <- oldseed
> rnorm(3, 100, 15)
[1] 107.21030  79.72381 101.61939
> rnorm(3, 100, 15)
[1] 85.63351 89.31402 94.36237
```

It should be noted that the random normal deviates sampled from the execution of the similar command lines may not be the same as those shown in the above example. Instead of saving and reusing the computer generated random number seed, the set.seed function can also be used to control the random sampling (e.g., type in set.seed(1234567) before using a random sample function).

R has many probability distributions to use for constructing the four types of functions. Their names and the parameters that should be used in the ordered arguments with the default values after the equal signs are as follows:

beta(shape1, shape2)	beta
binom(size, prob)	binomial
cauchy(location=0, scale=1)	Cauchy
chisq(df)	chi-square
exp(rate)	exponential
f(df1, df2)	Snedecor's F
gamma(shape)	gamma
geom(prob)	geometric
hyper(m, n, k)	hypergeometric
lnorm(meanlog=0, sdlog=1)	lognormal
logis(location=0, scale=1)	logistic
nbinom(size, prob)	negative binomial
norm(mean=0, sd=1)	normal
pois(lambda)	Poisson

```
t(df)                    Student's t
unif(min=0, max=1)       uniform
weibull(shape)           Weibull
wilcox(m, n)             Wilcoxon rank sum
```

An additional argument of a non-centrality parameter ncp can be used for the beta, chi-square, *F*, and *t* distributions. Using these functions, R virtually eliminates the need of nearly all statistical tables in the elementary, applied statistical textbooks.

The sample function takes a vector and the sample size for its arguments to perform random sampling. For example:

```
> x <- rnorm(4); x
[1] -0.2941517  0.7430328 -0.7172012 -1.1930889
> sample(x, 2)
[1] 0.7430328 -0.1712012
```

A.2.6 Inferential Statistical Methods

A number of functions are available to perform tests of simple statistical hypotheses. The one-sample *t* test, for example, can be performed:

```
> iq <- c(110, 120, 120, 130)
> t.test(iq, mu=100)

        One Sample t-test

data:  iq
t = 4.899, df = 3, p-value = 0.01628
alternative hypothesis: true mean is not equal to 100
95 percent confidence interval:
 107.0077 132.9923
sample estimates:
mean of x
      120
```

The independent two-sample *t* test with the equal variances assumption (i.e., the pooled two-sample *t* test) can be performed, for example:

```
> x1 <- c(6.5, 6.8, 7.1, 7.3, 10.2)
> x2 <- c(5.8, 5.8, 5.9, 6.0, 6.0,
+ 6.0, 6.3, 6.3, 6.4, 6.5, 6.5)
> t.test(x1, x2, var.equal=T)
```

```
        Two Sample t-test

data:   x1 and x2
t = 3.22, df = 14, p-value = 0.00617
alternative hypothesis: true difference in means is
  not equal to 0
95 percent confidence interval:
 0.4820473 2.4052255
sample estimates:
mean of x mean of y
 7.580000  6.136364
```

The independent two-sample *t* test without the equal variances assumption can be performed by not employing the var.equal argument.

To change the default confidence level of 0.95, for example, the conf.level = 0.99 argument can be used. For the *t*-test and for other relevant hypothesis tests, the alternative argument can be used to change the default two-sided alternative hypothesis into the one-sided alternative hypothesis. The default alternative = two.sided argument can be modified to either alternative = less or alternative = greater.

In addition to the t.test function, many other simple statistical inferences can be performed using the following functions:

binom.test	exact test for binomial proportions
chisq.test	chi-square test for a contingency table
cor.test	test of a correlation
fisher.text	Fisher's exact test for contingency tables
friedman.test	Friedman's rank sum test for randomized blocks
kruskal.test	Kruskal-Wallis rank sum test
ks.test	Kolmogorov-Smirnov test
mantelhaen.test	Mantel-Haenszel chi-square test
mcnemar.test	McNemar's chi-square test
prop.test	test for proportions
shapiro.test	Shapiro-Wilk normality test
var.test	folded *F* test to compare two variances
wilcox.test	Wilcoxon rank sum and signed rank sum tests

A.2.7 Statistical Models

Univariate statistical models try to explain or predict the value of one variable from the values of other variables. The modeling techniques that are available in R include:

lm	linear model
aov	analysis of variance
glm	generalized linear model
gam	generalized additive model
loess	local regression model
tree	tree-based model
nls	nonlinear model
ms	optimization

It is definitely beyond the scope of this appendix to cover the statistical background for the methods. The online help files and other statistical textbooks should be consulted for further information (e.g., Everitt and Hothorn 2010; Kabacoff 2011).

The analysis of variance can be performed with the aov function, then the summary function is used to obtain nicely formatted output:

```
> y <- c(1, 2, 2, 3, 2, 3, 3, 4, 3, 4, 4, 5)
> x <- c(1, 1, 1, 1, 2, 2, 2, 2, 3, 3, 3, 3)
> output <- aov(y ~ x)
> summary(output)
            Df Sum Sq Mean Sq F value  Pr(>F)
x            1      8     8.0   13.33 0.00445 **
Residuals   10      6     0.6
---
Signif. codes:  0 '***' 0.001 '**' 0.01 '*' 0.05 '.' 0.1 ' ' 1
> boxplot(y ~ x)
```

The aov function alone can generate a brief summary of the analysis of variance results. Note that the last command line is to get the side-by-side boxplot of the variable being compared.

The simple regression can be performed and detailed results can be displayed with the summary function:

```
> x <- c(1, 2, 2, 3)
> y <- c(1, 3, 2, 2)
> lm(y ~ x)

Call:
lm(formula = y ~ x)

Coefficients:
(Intercept)            x
        1.0          0.5

> output <- lm(y ~ x)
> summary(output)
```

A.2 R Basics

```
Call:
lm(formula = y ~ x)

Residuals:
         1          2          3          4
-5.000e-01  1.000e+00 -2.776e-17 -5.000e-01

Coefficients:
            Estimate Std. Error t value Pr(>|t|)
(Intercept)   1.0000     1.2990   0.770    0.522
x             0.5000     0.6124   0.816    0.500

Residual standard error: 0.866 on 2 degrees of freedom
Multiple R-squared:   0.25,     Adjusted R-squared:  -0.125
F-statistic: 0.6667 on 1 and 2 DF,  p-value: 0.5
```

The following three command lines yield the scatterplot of variables x and y, the available objects within the output, and the residual plot, respectively.

```
> plot(x, y)
> names(output)
 [1] "coefficients"  "residuals"   "effects"  "rank"
 [5] "fitted.values" "assign"      "qr"       "df.residual"
 [9] "xlevels"       "call"        "terms"    "model"
> plot(output$fitted.values, output$residuals)
```

In multiple regression with x1 and x2 as predictors, after defining the two variables, x1 and x2, the line can be used to obtain the regression results is:

```
> summary(lm(y ~ x1 + x2))
```

The followings are the most commonly used generic functions for extracting model information:

anova	ANOVA table for model comparisons
coef	extract the regression coefficients
deviance	residual sum of squares
formula	extract the model formula
plot	produce plots
predict	predicted values from the same set of variables
print	print a concise version of the object
residuals	extract the residuals
step	select a suitable model from an information index
summary	print a comprehensive summary of the results
vcov	returns the covariance matrix of estimates

In addition, the followings are multivariate statistical procedures in R:

cancor	canonical correlation
cmdscale	multidimensional scaling
discr	linear discriminant analysis
dist	calculating a distance matrix
hclust	hierarchical clustering
clorder	reorder a cluster tree
cutree	forming groups from a cluster analysis
labclust	labeling a cluster plot
leaps	all subsets regression using leaps and bounds
plclust	producing a cluster dendogram plot
subtree	extracting part of a cluster tree
prcomp	principal components analysis

A.2.8 Graphs

Many univariate, bivariate, and multivariate graphs are available in R. It should noted that there are also numerous options and arguments that can be employed in generating graphs and figures. Note that not all the functions will generate proper results. The following command dealing with trivial data, for example, can produce not really a reasonable figure (n.b., the last two command lines will generate proper histograms):

```
> x <- c(1,2,2,3)
> hist(x)
> hist(x, seq(.5,3.5,1))
> hist(x, seq(.5,3.5,1), xlim=c(0,4))
```

There are three groups of plotting commands in R: (1) high-level plotting functions to create a new plot on the graphics device, (2) low-level plotting functions to add more information to an existing plot, and (3) interactive graphics functions to interactively modify information from an existing plot using a mouse.

The followings are mostly the first group of the procedures to create graphs in R:

plot	generic function with many objects
boxplot	box and whisker plots
matplot	plots of two or more vectors of an equal length
qqnorm	quantile–quantile plots
qqline	draw a line in qqnorm
qqplot	distribution comparison plots
pairs	plots of a matrix or data frame
coplot	conditioning plots for 3 or more variables

A.2 R Basics

`dotchart`	construct a dotchart
`image`	plot of three variables
`coutour`	plot of three variables
`persp`	plot of three variables
`stem`	stem and leaf plot

For categorical variables, the following graphical summary can be used:

`barplot`	bar graph
`pie`	pie chart
`dotchart`	frequency summary with dots

Note that the plot command is a general graphical command, and there are numerous options and arguments to be used. To fully explore the command,

```
> help(plot)
```

or

```
> ?plot
```

will open up the manual pages of the plot command. One of the best ways to learn how to use the plot command is by executing the example command lines from the help pages. You can certainly copy and paste the example command lines from the manual page to a new script window (i.e., R Editor), modify portions, and execute the portions of the syntax by highlighting them and clicking the execution icon (i.e., the Run line or selection icon). In fact, the example command lines contained in the manual pages can be executed with the example function. For example, for boxplot:

```
> example(boxplot)
```

will show many examples from the manual pages. To learn how to use the graphical commands in R, remember that a picture is worth ten thousand words and to see is to believe. The best way to view and to perform some experiments of your own you can follow the procedure mentioned earlier; starting first with opening of the help page of the respective command and sequentially executing the example command lines with some of your own modifications. For a demonstration purpose, you may also obtain some R graphics capabilities by typing:

```
> demo(graphics)
```

Some plots are used with statistical modeling commands. The regression diagnostic plots, for example, can be obtained:

```
> x <- rnorm(100)
> y <- rnorm(100)
> par(mfrow = c(2,2))
> plot(lm(y ~ x))
```

The high level graphic commands will produce a plot which replaces the previous one. To open a new graphical output window instead of replacing the previous one, the following command line can be used before typing in a new high level graphic command:

```
> win.graph()
```

The followings are some examples of the command lines to list the devices, to identify the currently active device, to set a specific one as the current device, to close the current device, and to close all the graphics devices:

```
> dev.list()
> dev.cur()
> dev.set()
> dev.off()
> graphics.off()
```

The plot contained in the active R graphics device window can be copied or saved by right clicking the window. In terms of the saving the content in the R graphics device window the following seven functions of the respective file types are available (n.b., the parentheses in the end of the respective file types contain the size of the file obtained from the example runs; i.e., xyplot with respective extensions of eps, pdf, wmf, png, jpg, bmp, and tif):

`postscript`	PostScript or Encapsulated PostScript (7KB)
`pdf`	Portable Document Format (5KB)
`win.metafile`	Windows Metafile (15KB)
`png`	Portable network Graphics (3KB)
`jpeg`	Joint Photographic Experts Group (10KB)
`bmp`	Bitmap (227KB)
`tiff`	Tagged Image File Format (676KB)

You can save a resulting plot, for example, by typing in the following command lines:

```
> x <- c(1, 2, 3, 2); y <- c(1, 3, 2, 2)
> postscript("xyplot.eps")
> plot(x, y)
> dev.off()
```

As a real example, Fig. 1.1 in Chap. 1 used the following command lines:

```
> postscript("BIRTFigure1p1.eps", width=3.5, height=2.5,
    pointsize=7)
> par(lab=c(7,3,3))
> theta <- seq(-3, 3, .1)
> b <- 0
> a <- 1
> P <- 1 / (1 + exp(-a * (theta - b)))
> plot(theta, P, type="l", xlim=c(-3,3), ylim=c(0,1),
    xlab=expression(paste("Ability, ",theta)),
```

A.2 R Basics

```
        ylab=expression(paste(
        "Probability of a Correct Response, P(",theta,")")))
> dev.off()
```

The LaTeX input lines, if you know what this is about, are as follows:

```
\begin{figure}[ht]
\centering
\includegraphics[scale=0.3,angle=-90]{BIRTFigure1p1.eps}
\caption{A typical item characteristic curve.}
\label{BIRTFigure1p1}
\end{figure}
```

A.2.9 *Missing Values*

In R, not available (i.e., NA) is used as a missing value. The following lines show how the missing values are treated in R:

```
> x <- c(1, NA, 3, 2); x
[1]  1 NA  3  2
> is.na(x)
[1] FALSE  TRUE FALSE FALSE
> sum(!is.na(x))
[1] 3
> newx <- x[!is.na(x)]; newx
[1] 1 3 2
> x[2] <- sum(newx)/sum(!is.na(x)); x
[1] 1 2 2 3
```

Note also that NaN (i.e., not a number) and Inf (i.e., infinity) are treated as missing cases.

```
> x1 <- 0/0
> x2 <- Inf
> x3 <- Inf - Inf
> x <- c(x1, x2, x3, 2); x
[1] NaN NaN NaN   2
> is.na(x)
[1]  TRUE  TRUE  TRUE FALSE
```

The best way to solve the problem of missing values is prevention of the occurrence of missing in the data collection process. There is no missing strategy, none whatsoever, how sophisticate and complicate it can be, that is better than obtaining complete data. Obviously, there is no royal road for missing.

A.3 R Packages

As many general purpose statistical computer programs, R consists of packages that are physically stored in the library of R. A package contains a set of R functions, data, and compiled code. The command lines that can be used to locate the library and the list of packages available are as follows:

```
> .libPaths()
> library()
```

The packages can be installed and loaded with the command lines. For example, the MASS package can be accessible by:

```
> install.packages("MASS")
> library(MASS)
```

The R package can be loaded from the web site:

http://cran.r-project.org/web/packages

Note that you can also use the Packages button and accompanied options in the R console window to perform the same operations.

The standard packages include base, datasets, utils, grDevices, graphics, stats, and methods. There are numerous special packages you can download and install. For example, the R packages for item response theory modeling include ltm (Rizopoulos 2006), eRm (Mair and Hatzinger 2007), and mirt (Chalmers 2012). The use of other than the standard packages has been largely avoided in this appendix.

Appendix A References

Abramowitz, M., & Stegun, I. A. (Eds.). (1964). *Handbook or mathematical functions with formulas, graphs, and mathematical tables.* Washington, DC: U.S. Government Printing Office.

Becker, R. A., & Chambers, J. M. (1984). *S: An interactive environment for data analysis and graphics.* Belmont, CA: Wadsworth.

Becker, R. A., & Chambers, J. M. (1985). *Extending the S system.* Monterey, CA: Wadsworth.

Becker, R. A., Chambers, J. M., & Wilks, A. R. (1988). *The new S language: A programming environment for data analysis and graphics.* Pacific Grove, CA: Wadsworth.

Chalmers, R. P. (2012). mirt: A multidimensional item response theory package for the R environment. *Journal of Statistical Software, 48*(6), 1–29.

Chambers, J. M. (1998). *Programming with data: A guide to the S language.* New York, NY: Springer.

Chambers, J. M., & Hastie, T. J. (Eds.). (1993). *Statistical models in S.* New York, NY: Chapman & Hall.

Everitt, B. S., & Hothorn, T. (2010). *A handbook of statistical analyses using R.* Boca Raton, FL: Chapman & Hall/CRC.

Kabacoff, R. I. (2011). *R in action: Data analysis and graphics with R.* Shelter Island, NY: Manning.

Mair, P., & Hatzinger, R. (2007). Extended Rasch modeling: The eRm package for the application of IRT models in R. *Journal of Statistical Software, 20*(9), 1–20.

A References

Rizopoulos, D. (2006). ltm: An R package for latent variable modeling and item response theory analyses. *Journal of Statistical Software, 17*(5), 1–25.

Spector, P. (1994). *An introduction to S and S-PLUS*. Belmont, CA: Duxbury Press.

Venables, W. N., Smith, D. M., & The R Development Core Team. (2009). *An introduction to R* (2nd ed.). LaVergne, TN: Network Theory.

Appendix B
Estimating Item Parameters Under the Two-Parameter Model with Logistic Regression

In Chap. 3 the estimation of item parameters by fitting an item characteristic curve to the observed proportions of correct response was illustrated by generating item parameters and the data for the 33 groups of examinees along the ability scale. For each group there were 21 examinees. As mentioned in the section of maximum likelihood estimation of item parameters, the actual computation to estimate item parameters using the Newton-Raphson method will be complicated even though the ability scores are assumed to be known. For the two-parameter logistic model, however, it is possible to use logistic regression via an R function glm (i.e., generalized linear model). The following example illustrates how to use such an R function to obtain the estimates of the item difficulty and item discrimination parameters. It should be noted that the method used in the R function of glm is the iterative reweighted least squares method instead of the method of maximum likelihood.

The data generated and used in Fig. 3.1 will be used in this example. In Fig. 3.1, 33 observed proportions of correct response were plotted along the ability scale. The length of the theta, ability vector was 33. The observed proportions were based on the random variates generated from the binomial distribution. For each ability score level, there were 21 examinees yielding $f_g = 21$. The probability of correct response $P(\theta_g)$ was obtained with the generated item discrimination parameter $a = 1.27$ and the generated item difficulty parameter $b = -0.39$. Before obtaining the vector p of the observed proportions of correct response, the function rbinom(length(theta), f, P) randomly generated the vector that contains the numbers of examinees with correct response. Assume that we have saved such a vector by assigning it a name r under the two-parameter model using, for example, the following command lines:

```
> theta <- seq(-3, 3, .1875)
> f <- rep(21, length(theta))
> wb <- -0.39
> wa <- 1.27
```

```
> for (g in 1:length(theta)) {
    P <- 1 / (1 + exp(-wa * (theta - wb)))
  }
> r <- rbinom(length(theta), f, P)
```

For the data presented in Fig. 3.1, r contained the following 33 numbers:

1, 0, 0, 1, 4, 1, 0, 5, 1, 5, 3, 3, 11, 11, 7, 13, 12, 15, 17, 15, 15, 17, 20, 19, 19, 21, 18, 19, 21, 20, 21, 21, 21

The vector p in the plot was obtained, in essence, by dividing it with f:

```
> p <- r / f
```

In logistic regression, the probability of correct response will be modeled as a function of known ability score levels as:

$$P(\theta_g) = \frac{1}{1 + \exp[-(\beta_0 + \beta_1 \theta_g)]}, \quad (B.1)$$

where β_0 and β_1 are the intercept parameter and the weight parameter on the ability score variable, respectively. Estimates of these parameters are expressed as b_0 and b_1. It is clear that under the two-parameter logistic model, $a \equiv \beta_1$ and $b \equiv -\beta_0/\beta_1$.

Assume that we have the vectors of r and f as well as theta. The following R command lines will be used to obtain the logistic regression results:

```
> glm(cbind(r,f-r) ~ theta, family=binomial)
```

The obtained results show that $b_0 = 0.386$ and $b_1 = 1.377$. Instead of calculating the corresponding values of item discrimination and item difficulty parameter estimates from these values reported in the results from glm, the following command lines can be used because the estimates may have more than three decimal places:

```
> lroutput <- glm(cbind(r,f-r) ~ theta, family=binomial)
> ahat <- summary(lroutput)$coefficients[2]
> bhat <- -summary(lroutput)$coefficients[1] /
    summary(lroutput)$coefficients[2]
> ahat; bhat
```

The estimates of item parameters are $\hat{a} = 1.377101$ and $\hat{b} = -0.2802905$. These estimates are very close to the corresponding item parameters used to generate the observations, that is, $a = 1.27$ and $b = -0.39$.

Note that elements or portions of the output of the R function (e.g., lroutput) can be seen with the function names, for example:

```
> names(lroutput)
```

The elements shown can be extracted with the dollar sign $. If only the numerical values are needed, then we may use the function summary as shown in the above command lines.